Yanmar

YANMAR DIESEL ENGINE MODEL 2 S

Service Manual

Yanmar

YANMAR DIESEL ENGINE MODEL 2 S

Service Manual

ISBN/EAN: 9783954272754
Erscheinungsjahr: 2013
Erscheinungsort: Bremen, Deutschland

© maritimepress in Europäischer Hochschulverlag GmbH & Co. KG, Fahrenheitstr. 1, 28359 Bremen. Alle Rechte beim Verlag und bei den jeweiligen Lizenzgebern.

www.maritimepress.de | office@maritimepress.de

Bei diesem Titel handelt es sich um den Nachdruck eines historischen, lange vergriffenen Buches. Da elektronische Druckvorlagen für diese Titel nicht existieren, musste auf alte Vorlagen zurückgegriffen werden. Hieraus zwangsläufig resultierende Qualitätsverluste bitten wir zu entschuldigen.

YANMAR DIESEL ENGINE
SERVICE MANUAL | MODEL 2S

CONTENTS

PREFACE
1. OUTLINE OF MAJOR STRUCTURES 1
2. FUEL AND LUBRICATING OIL 2
   2-1. Fuel 2
   2-1-1. Light Oil 2
   2-1-2. Heavy Oil 3
   2-1-3. Quality Fuel Oil 4
   2-1-4. Properties of Fuel & Engine Performances 8
   2-1-5. Cautions on Fuel 9
   2-2. Lubricating Oil 9
   2-2-1. Engine Oil 9
   2-2-2. W & Multi Purpose Grade Oils 11
   2-2-3. Property Requirements for Engine Oil 12
   2-2-4. Exchange of Engine Oil 12
   2-2-5. Cautions for Handling Engine Oil 13
3. PERIODICAL CHECKING & SERVICING 15
   3-1. Periodical List of Items to be Checked and
        Frequency of Checks 15
   3-2. Checking Locations 20
   3-3. Checking Locations by Systems 20
4. MAINTENANCE STANDARDS OF MAIN PARTS 22
   4-1. Maintenance Standards 22
   4-2. List of Measuring Positions 23
   4-3. List of Wear Limit 26
   4-4. List of Undersize Metals 27
   4-5. List of Oversize Metals 27
5. ENGINE DISASSEMBLY 28
   5-1. Precautions Prior to the Disassembly of the Engine ... 28
   5-2. General Cautions for Maintenance and Cleaning 30
   5-3. Cautions for Disassembling the Engine 30
   5-4. Procedure on Disassembling 31
6. ENGINE REASSEMBLY 47
   6-1. Cautions for Reassembling the Engine 47
   6-2. Procedure on Reassembling 47
7. DISASSEMBLY & REASSEMBLY OF OTHER PARTS 65
   7-1. Fuel Injection Pump 65
   7-2. Adjustment of Governor 67
   7-3. Fuel Injection Time 69
   7-4. Fuel Injection Valve 71
   7-5. Air Venting of Fuel Injection System 73
   7-6. Cylinder Head 74
   7-7. Piston Pin and Connecting Rod 79

7-8.	Cylinder Liner	81
7-9.	Replacement of Crank Journal Metal	81
7-10.	Camshaft and Camshaft Mountings	82
7-11.	Cooling Water Pump	82
7-12.	Lubricating Oil Pump and Lubricating Oil Pressure Adjusting Valve	83
8.	COUNTERMEASURES TO ENGINE TROUBLES	87
9.	STORING ENGINE	89

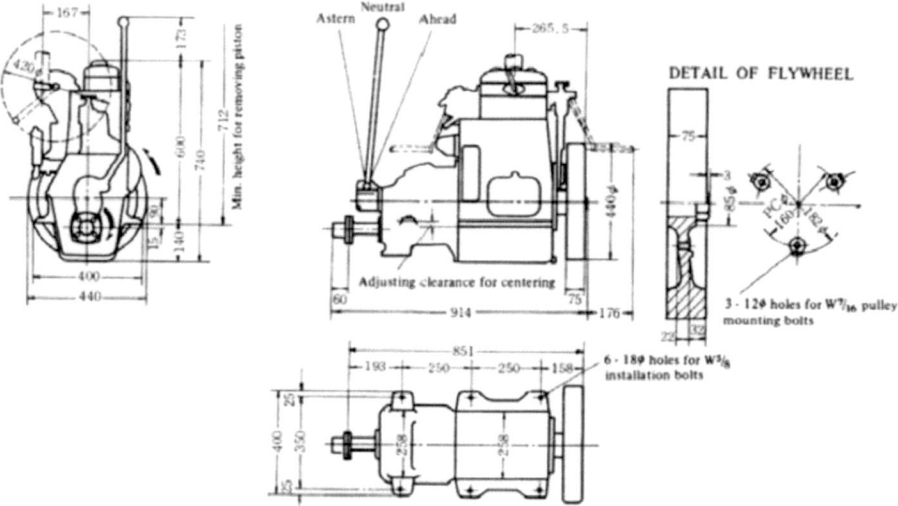

This figure shows installation details to be viewed from the engine bottom.

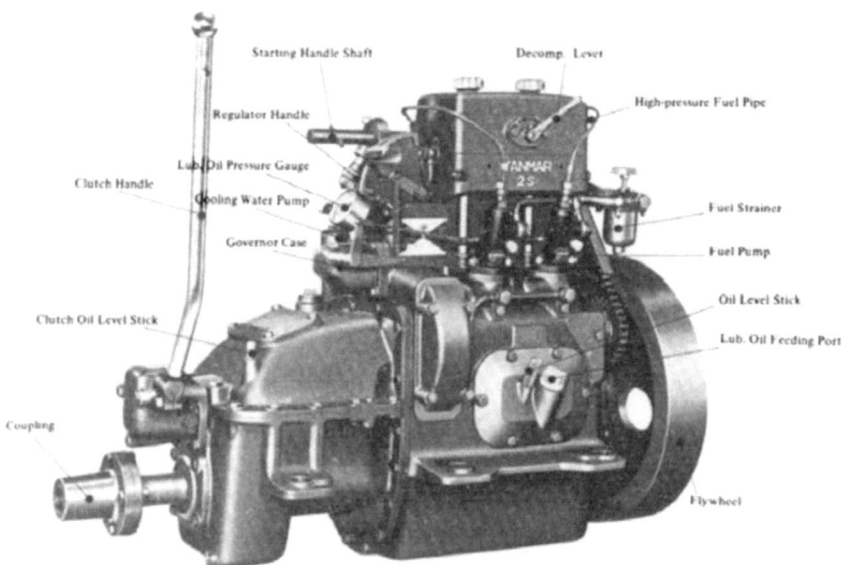

PREFACE

To operate the engine always under its best conditions, remember the following five points:

* The engine always requires clean fuel oil of the best quality.
  Always keep the fuel tank, strainers, and fuel pipes clean.
* The engine is always in need of clean lubricating oil of the best quality.
  Use an adequate grade lubricating oil and maintain the oil level above the minimum line on the oil level gauge at all times.
* The engine must always have clean air.
  Check that there are no carbon deposits or other foreign particles precipitated at the air intake holes and exhaust system.
* The engine must always be water cooled.
  Supply a sufficient amount of cooling water in the cooling system.
* The engine functions more efficiently under normal load conditions.

1. OUTLINE OF MAJOR STRUCTURES

This outline briefly describes the structure of engine by the following 10 major groups.

No.	Major Group	Part	Description of Structure
1.	Engine main body section	Cyl. block	Monoblock cast of water jacket, crank case and oil sump
		Cyl. liner	Separate piece from cyl., wet type, chrome plated
		Main bearing	KJ triple round metal
		Oil sump	Monoblock cast with cyl. block
2.	Suc./exh. device & valve drive mechanism	Cyl. head	Monoblock cast of two cylinder, water-cooled
		Suc./exh. valve	Mushroom valve (120°)
		Exhaust silencer	No-resistance type
		Valve drive mechanism	Hardened, polished sliding contact portion of cam, tappet, valve arm
3.	Main working section	Crankshaft	Stamp forging, hardened, polished of pin journal
		Flywheel	Mounted with crankshaft and taper part
		Piston	Trunk piston
		Piston ring	3 compression rings & 2 oil scraper rings
		Piston pin	Wholly floating type
		Connecting rod	I-shaped main portion, stamp forging
		Crankpin metal	KJ triple thin metal
4.	Lub. oil system	Lub. oil pump	Spur gear type
		Lub. oil strainer	Pump suction side, hole bored steel plate; Delivery side, auto-clean type
		Oil level checking device	Oil level gauge
		Lub. oil cooler	Fin, full-capacity passing type
5.	Cooling system & bilge system	Cooling water pump	Reciprocating plunger type pump
		Bilge pump	Reciprocating plunger type pump (special order)
6.	Fuel system	Fuel injection pump	Bosch type, individual per cylinder
		Fuel injection valve	Sealed, automatic valve, pintle type
		Fuel oil strainer	Auto-clean type
7.	Governor system	Governor	Centrifugal, all-speed control type
8.	Starting system	Chain starting	Multiplying starting by chain (Electric starting as special order)
9.	Power transfer mechanism	Speed reduction gear	Spur gear type with built-in reversing clutch
		Reversing clutch	Single-plate disc. clutch mechanism
10.	Instruments	Tachometer	
		Pressure gauge	Bourdon tube type

2. FUEL AND LUBRICATING OIL

As you already know that there are many classes and grades of fuel and lubricating oil marketed today for use in diesel engines. Naturally, a selection of wrong class and/or grade of fuel or lubricating oil might result in unexpected trouble of a diesel engine or otherwise sure to shorten the serviceable life of the engine. Use of quality fuel and lubricating oil of right class and grade will increase the life of the engine many times, offsetting the higher price of the quality oil because on long terms long serivce of the engine gives its owner much more savings than use of low-cost oil which tends to shorten the engine life.

2-1. Fuel

Except gas engine, nearly all types of internal combustion engine burn fuel derived from petroleum for power source. In the following paragraphes, we will explain you light oil and heavy oil used as fuel to run diesel engines.

2-1-1. Light Oil
1. Diesel Light Oil
   In general, light oil has the specific gravity of 0.83~0.89 and the boiling point of 200°C~370°C. Diesel light oil is widely used to run high-speed diesel engines of 1200 rpm or mor employed in agricultural machinery, automobile, construction equipment, etc.
2. Requirements of Diesel Light Oil
   1) High cetane rating:
      Good ignitability and high combustion efficiency. Generally these requirements are met diesel light oil of the cetane number of over 45.
   2) Low sulphur content:
      High sulphur content of the oil speeds up corrosion and wear of engine parts particularly those parts which directly come in contact with fuel. Because of these reasons, such oil should not contain 1% or more of sulphur.
   3) Appropriate viscosity:
      Degree of viscosity must be appropriate with relation to ignition and combustion. If the viscosity is too high, atomized fuel particles are too large for dispersion; thus, the combustion time lags and color of the exhaust gas becomes poor. On the other hand, if the viscosity is too low, the atomized particles are too small for penetration of injection, resulted in scorching of the plunger and injection nozzle as they are not provided with lubricating action.
   4) No mixed dust and moisture:
      Impure oil usually contains dust and moisture which cause damage to the plunger and injection nozzle. We recommend use of pure fuel. Besides, be sure to filter the fuel prior to supply to the engine.

3. Cetane Rating

Cetane rating is the most conveniently used criterion for rating diesel fuel and is equivalent of octane rating for gasoline. Cetane rating is used as index of the ignitability and refractoriness. Low cetane number fuel has poor ignitability and tends to cause diesel knock. Causes of diesel knocking are in many cases opposite to those of knocking of gasoline engine, as you can see in the following table.

Causes of Engine Knocking

		Gasoline Engine	Diesel Engine
Engine	Compression Ratio	High	Low
	Temperature & Pressure of Suction Air	High	Low
	Temperature of Cylinder Wall	High	Low
	RPM	Low	High
	Ignition Point of Fuel	Low	High
	Cylinder Capacity	Large	Small
Fuel	Octane Number	Low	———
	Cetane Number	———	Low

2-1-2. Heavy Oil

1. Diesel Heavy Oil

    Diesel heavy oil is generally used to run low-speed (average piston speed is below 5 m/sec), intermediate-speed (average piston speed is 5~6 m/sec) and high-speed (average piston speed is over 6 m/sec) diesel engines employed to power large marine vessels.

2. Requirements of Diesel Heavy Oil

    1) Low viscosity:

    In general, low viscosity is preferable because such fuel is easily injected and consequently combusted well.

    B grade heavy oil is used for low-speed, large diesel engines; C grade oil is also used in the large marine vessel engine (of low speed) equipped with heater from the point of view of cost consideration.

    2) Low ashes content:

    For same reasons as described for sulphur content in connection with diesel light oil, ashes content should be trace amount.

    In additions, other properties of diesel light oil also apply for diesel heavy oil generally.

    3) Low pour point:

    In case the engine is not provided with heater, it is convenient to use a heavy oil having a low pour point.

    4) Low residual carbon and sulphur contents:

    These requirements are also met in regard to same reason given for low sulphur content of diesel light oil.

3. Classification of Heavy Oil

Grade of Heavy Oil	Specific Gravity	Viscosity (50°C) (Redwood)	Application
A	about 0.85~0.89	<85 sec.	diesel engine
B	about 0.90~0.935	about 85~200 sec.	diesel engine
C	about 0.93~0.96	about 200~600 sec.	heater type engine 6

For fuel to be used, the followings which best suitable for the engine are recommended. If, however, inferior fuel is used, poor exhaust color, damage of exhaust valve and its seat, fuel pump, fuel injection valve, and early abrasion of cylinder liner, piston and piston rings may be resulted. Use as good-quality fuel oil as possible.

Supplier	Brand Name
SHELL	Sheel Diesoline or local equivalent
CALTEX	Caltex diesel oil
MOBIL	Mobil diesel oil
ESSO	Esso diesel oil

2-1-3. Quality Fuel Oil

Most suitable oil as fuel for Yanmar Diesel Engines should possess the properties such as specific gravity, ignition point and viscosity similar to light oil.

It, along with diesel light oil, is widely used to power high-speed diesel engines.

1. Features
    1) Good fuel injection performance and lubricating action on fuel injection pump:
       Combustion of fuel in diesel engine begins with the injection nozzle. For the atomization, low viscosity requirement is prerequisite. However, to combust completely the atomized fuel, appropriate penetration of injection is rerquired of fuel in mist form. In other words, to combust the fuel well, it is necessary to provide the atomized fuel to mix well with air uniformly throughout the combustion chamber instead of the fuel distributed only in neighborhood of the injection nozzle. Thus, to give fuel some degree of penetration of injection, in this respect, the viscosity of fuel must be maintained at sufficient level. Limited by these two factors, the range of appropriate viscosity is determined. Such viscosity appropriate for fuel of small-size, intermediate/high-speed engines is less than 89.5 seconds for RW (Redwood) No. 1, 50°C, as the experimental result.

       Furthermore, inside the fuel injection pump, both the cylinder and plunger are to make reciprocating motion violently through the high precision processed narrow space. To provide sufficient lubrication on these parts, fuel itself must provide this additional function. Therefore, the viscosity of fuel must be such to prevent abnormal wear and scorching of the plunger. For this regard, experiment revealed that such viscosity value is more than 32 seconds with RW (Redwood) No. 1, 50°C. From these two findings, the sufficient viscosity required of diesel fuel is within 32~89.5 seconds with RW No.1, 50°C Yet for high-speed diesel engines, the range of appropriate viscosity is in neighborhood of 32~38 seconds. Fuel oil should be produced with consideration given to above points; thus, it provides excellent atomization and good lubricating action on fuel injection pump.

2) Good ignitability and no diesel knock:

In diesel engine, it is ideal to have combustion occurring immediately upon ignition of the fuel injected into high-temperature compressed air. In detail, the injected fuel particles absorb heat from the surrounding air and begin to evaporate from their surfaces. Finally fuel vapors permeate through air until spontaneous ignition temperature is reached. At such time, ignition initially occurs. Time required for ignition from the moment fuel is injected is called "ignition waiting period" or "ignition delay." If ignition delay is significant, the fuel injected prior to ignition remains as it is and, when ignition initiates, begins to combust at once violently, raising the combustion pressure abruptly. Related to this combustion pressure rise, the pressure wave forms to knock the piston and cylinder wall as high metallic sound is heard. This is what is commonly known as "djesel knock" or "combustion knock" and links to poor run or out of order of engine. Thus, good ignitability of fuel is the most important factor for high-speed diesel engines. Ignitability is measured by crank angle, time required to ignite from the moment fuel is injected into the combustion chamber. However, commonly used measuring stick for ignitability of fuel itself is cetane rating. Higher the cetane number, better the ignitability of the fuel. Yet when considered from all angles, we cannot say that higher cetane number means better ignitability; there is appropriate value of cetane rating with relation to engine type and rpm. We may list below one view on this matter.

	Cetane Number
High-speed Diesel Engine	40 ~ 55
Intermediate-speed Diesel Engine	30 ~ 40
Low-speed Diesel Engine	20 ~ 30

Naturally, high cetane rating fuel is not required for intermediate- and low-speed engines. But in high-speed (greater than 1000 rpm) engine, inappropriate cetane rating fuel affects considerably engine performance. If fuel of lower than proper cetane rating is used, the following troubles may take place.

a. Difficult start.
b. Poor operation enven started.
c. High combustion pressure, causing diesel knock.
d. If diesel knocking occurred, output drop and seizure of engine due to overheat.
e. Damage of fuel injection nozzle and exhaust valve.
f. Intense smoking, increasing carbon accumulation inside engine and consequently soiling of oil.
g. Deterioration of oil, speeding up wear of piston rings, ring grooves, cylinder liner, etc.

It is advisable to use fuel oil with cetane rating of about 55; thus its ignitablity is superior, start of engine is easy with it, and no diesel knock is resulted.

Relation between Cetane Rating and Starting Characteristics

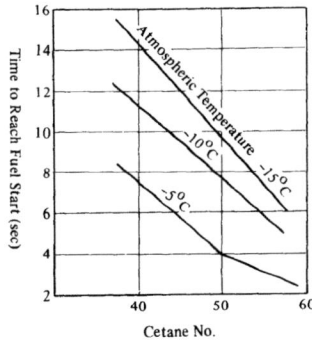

Relation between Cetane Rating and Ignition Delay

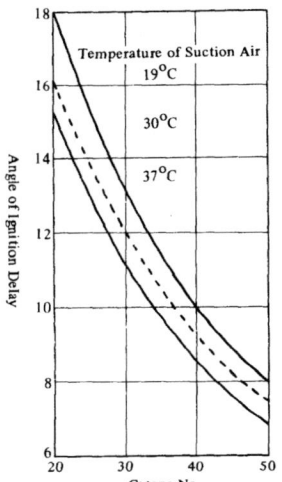

3) Minor corrosion effect

Major agent affecting corrosion of engine parts is sulphur compound contained in diesel fuel. Unlike gasoline and kerosene, diesel fuel is not treated to remove sulphur; thus, the fuel contains considerable amount of sulphur compound. When diesel fuel is combusted, its content of sulphur compound forms sulfurous acid gas and a certain portion of this gas further reacts with air to form sulfuric anhydride. This in turn combines with water vapor produced as a part of the fuel combustion and becomes sulfuric acid. If cooled below the freezing points, it is condensed and corrode the metallic parts of engine it came in contact with. Formation of sulfuric acid is shown by chain reaction as follows:

$$S + O_2 \longrightarrow SO_2$$
$$2SO_2 + O_2 \longrightarrow 2SO_3$$
$$SO_3 + H_2O \longrightarrow H_2SO_4$$

Cause of fast corrosion and wear-off of diesel engine parts is traced to the above chemical reactions in most cases. Particularly, when the engine is operated under a low atmospheric temperature and cooling water temperature of below 70°C, such undesirable effect takes place at a fast pace. Besides, during short-time running of the engine with light load or while the engine is at rest, this corrosion also tends to proceed. As the freezing point of the combusted gas is raised with the volume of sulphur content, corrosion of engine parts and soiling of oil become intense as the sulphur compound content increases.

Fuel oil should have been produced with sufficient consideration given to damaging effect of sulphur compound and contains extremely little of it, freeing engine parts from corrosion of this type.

4) No clogging of fuel injection nozzle

Fuel injection pump is very precise part of a diesel engine; thus, even a small foreign particle particularly small solid matter, might disable fuel injection mechanism. Among undesirable solid particles concerned here are (1) dust presented in air and (2) iron rust from oil drum or fuel tank entered into fuel. Naturally the latter one is in most cases filtered out by strainers before reaching the fuel injection system. However, for some resosns, if such particle penetrated to the injection system, it causes major troubles of abnormal wear-off of the fuel injection mechanism and of clogging of the nozzle. In addition, if moisture is mixed into fuel in noticeable degree, it affects the injection system adversely although indirectly, differing from directness of bad effect of the solid particle. In detail, the moisture hinders the filtering ability of fuel strainer and cause rust within the fuel system and on the nozzle during engine rest. Besides, it gives a bad effect on fuel injection and thus causes incomplete combustion consequently. Fuel oil should be given strict quality control throughout production and storage stages. Thus, foreign particle and moisture are near nil.

5) Economy

Output and fuel consumption rate of a diesel engine are largely influenced by the cetane number and specific weight of fuel used. Fuel of sufficient cetane number is completley combusted and is said to have a high efficiency of combustion. Thus, fuel consumption rate (gr/HP/hr) is low. On the other hand, if fuel of insufficient cetane number is used, incomplete combustion with smoking occurs, causing diesel knocks. Thus, such fuel has a low effective pressure, and naturally its consumption rate must be higher. Fuel consumption rate, however, is much more depended upon the specific weight (calorific value) than the cetane rating of the fuel. If the specific weight is low, the combustion rate (gr/HP) is raised abruptly.

6) Good for winter as well as summer

It is recommended to use such an oil having the pour point below $-15°C$; thus, even in severe winter season, it combusts well.

2-1-4. Properties of Fuel & Engine Performances

Property of Fuel	Starting Characteristic	Smoothness of Operation	Smoke Generation	Exhaust Fume	Output	Fuel Consumption	Accumulation within Combustion Chamber
Ignitability (Cetane Rating)	No direct relation but higher cetane rating, better starting characteristic.	No direct relation but higher rating, better smoothness of operation.	Closely related; not much difference if cetane rating is above the min. cetane number limit.	Direct relation; higher cetane rating, lower exhaust fume.	No relation.	No relation.	Relation exists; higher cetane rating, lower the accumulation.
Volatility (90% end point)	No clear relation.	Relation exists; poorer volatility, better the performance.	Direct relation; better volatility, higher smoke generation.	No clear relation.	No relation.	No relation.	Relation exists; smaller this property, larger this aspect.
Viscosity	No clear relation.	Relation in a certain degree, high viscosity is not good.	Relation exists; higher viscosity, dependant on relation with the higher volatility.	Not related specifically.	Not related.	Not related.	Direct relation exists; dependant on relation with volatility.
Specific Weight	Not related.	Not related.	Direct relation exists; dependant on relation with volatility.	Not related specifically.	Direct relation exists; dependant on relation with calorific value.	Direct relation exists; dependant on relation with calorific value.	Relation exists; dependant on engine characteristic.
90% Residual Oil & Carbon Content	Not related.	Not related.	Relation exists; slightly inverse relation.	Not related specifically.	Not related.	Not related.	Relation exists; slightly inverse relation.
ASTM Gum	Not related.	Not related.	Relation exists; slightly inverse relation.	Not related specifically.	Not related.	Not related.	Relation exists; slightly inverse relation.

Property of Fuel	Starting Characteristic	Smoothness of Operation	Smoke Generation	Exhaust Fume	Output	Fuel Consumption	Accumulation within Combustion Chamber
Sulphur Content				Not related specifically.			
Flash Point				Not related specifically.			

2-1-5. Cautions on Fuel

It is important to select the fuel which does not contain dust and water vapor particularly. Besides, prior to filling the fuel tank, filtering is required. To filter fuel, use a piece of clean, finely interwoven cotton or the like cloth. Fuel come in drum can contains some amount of impurities such as dust and water particles settled down on its bottom. Refrain from inverting the can prior to transferring fuel therefrom or avoid pumping up fuel from the very bottom of the can.

2-2. Lubricating Oil

Lubricating oils presently used for all types of internal engines are about all of them mineral oil refined from petroleum. Depending upon their application, viscosity and quality (superior, regular, with/without additive), there are numerous kinds marketed today. Among them are lubricants for diesel engine use. We shall describe the appropriate kinds, property requirements and handling method of this group of lubricants shortly.

2-2-1. Engine Oil

1. Purpose of Engine Oil Usage

   Engine oil mainly purports to check on friction and wear-off of clearance between the cylinder wall and piston rings and bearing section of the pin and journal portions of crankshaft. Besides, not only it seals off a gap between the cylinder wall and piston rings and thereby prevents blow-by of combustion gas and consequently rules out decrease of produced output, but also remove harmful impurity from various sections of the engine, playing a role of preventing corrosion and rusting as well as of carrying away and cooling the heat due to friction.

2. Types of Engine Oil

   Engine oil is roughly classified into two groups; namely, motor oil (gasoline engine oil) and diesel engine oil. Within respective group, they are further classified into several types, denpended upon quality, usage condition, and viscosity.

Classification by Usage Condition
(API Service Classification)

Classification	Symbol	Quality & Application
Gasoline Engine Use	ML	Used under the most favorable operational condition of gasoline engine. No particular requirement is called for its lubricating action as the amount of worn-off particles is small.
	MM	Used under the average or heavy degree in severity of operational condition of gasoline engine. Consideration given to counteract corrosion of bearing and producing of worn-off particles when temperature of oil inside the crankcase is high.
	MS	Used under the severe operational condition of gasoline engine. Special lubricating action is called for taking a measure against producing of worn-off particles and corrosion of abrased bearings from standpoint of designed operational condition of the engine.
Diesel Engine Use	DG	Used under the light-load operational condition of diesel engine. Requirement is small in checking on producing of worn-off particles and abrasion due to designed roles of fuel, lubricating oil and engine itself.
	DM	Used under the severe operational condition of diesel engine when producing of worn-off particles, abrasion, etc. are evident due to sulphur content of fuel used or when residual carbon of lubricating oil affects greatly upon engine design.
	DS	Used under the heavy-load operational condition of disel engine when the fuel producing worn-off particles to a great extent, or when abrasion of the engine especially designed for high-output, high-load operation, etc. is serious.

NOTE: *Symbols operated in above table have the following meanings.*

	Symbol	Meaning
First Letter	M	Motor
	D	Diesel
Second Letter	L	Light
	M	Moderate
	S	Severe
	G	General

Engine oil has been refined and blended with addition of such necessary additives as oxidation inhibiter, corrosion inhibiter, rust preventive, dispersant, etc.

Classification by Viscosity
(SAE Viscosity Classification)

There are seven SAE numbers; namely, #5W, #10W, #20W, #20, #30, #40 and #50.
As list in the following table, engine oils are classified on base of viscosity alone. However, this classification is used for many decades and throughout the world; thus SAE number can well stand for convenient index on quality of engine oil.

List of SAE Viscosity Classification

SAE No.	0°F (-17.8°C)		210°F (98.9°C)	
	Saybolt Universal Viscosity (sec)	Kinematic Viscosity (CSt)	Saybolt Universal Viscosity (sec)	Kinematic Viscosity (CSt)
5W	4,000	869	—	—
10W	* 6,000 ~ 12,000	* 1,303 ~ 2,606	—	—
20W	** 12,000 ~ 48,000	** 2,606 ~ 10,423	—	—
20	—	—	45 ~ 58	5.73 ~ 9.62
30	—	—	58 ~ 70	9.62 ~ 12.93
40	—	—	70 ~ 80	12.93 ~ 16.77
50	—	—	85 ~ 110	16.77 ~ 22.68

NOTE: * *If kinematic viscosity is greater than 4.18 CSt (saybolt universal viscosity of 40 sec) at 98.9°C, it is permissible to have the viscosity of less than 6,000 seconds at -17.8°C.*

** *If kinematic viscosity is greater than 5.73 CSt (saybolt universal viscosity of 45 sec) at 98.9°C, it is permissible to have the viscosity of less than 2,606 CSt (saybolt universal viscosity of 12,000 sec).*

Reference: *SAE (Society of Automotive Engineers) Handbook (1959).*

2-2-2. W & Multi-purpose Grade Oils

The table of SAE viscosity classification listed above specifies only viscosities at 210°F for #20~#50 oils and at 0°F for #5W~#20W oils. Thus, there is no guarantee against some other temperatures. (Those oils designated with number suffixed by letter "W" are specified at a low temperature; therefore, they are generally used during very cold weather.) As can be seen in a graph below, even A oil and B oil are #30 oil, better oil has higher viscosity index and a smaller temperature change like in case of the B oil.

On the other hand, in case of the C oil in the graph below, it has the specified viscosity of #10W oil and in addition that of #30 oil at 210°F. Such oil as the C oil is called a multi-grade oil; C oil is #10W-30 multi-grade oil. Besides, #10W-30 oil, there are other multi-grade oils like #5W-20 and #20W-40 oils. Among them, #10W-30 is the most suitable for use in YANMAR air-cooled diesel engines as well as for the water-cooled diesel engines to be operated in cold region.

Viscosity of Multi-grade Oil

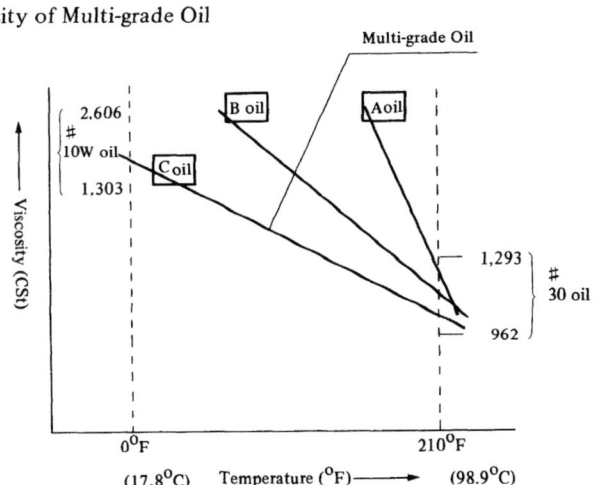

2-2-3. Property Requirements for Engine Oil

To operate an engine safely and economically, engine oil having the following properties is recommended.
1. Appropriate viscosity.
2. High viscosity index.
   Must have appropriate viscosity regardless of low or high temperature. (Make initial run easy and retention of oil film retainable.)
3. Low pour point. (Low solidifying point.)
4. Excellent oiliness, strong oil film, and good abrasion prevention. (Extends the lifetime of engine by preventing abrasion and seizure of metals.)
5. Large oxidation stability. (Extends the lifetime of engine as the oil does not deteriorate even subjected under worse operational condition.)
6. Corrosion inhibition. (Extends the lifetime of engine by preventing corrosion and rusting from forming on various parts of engine.)
7. Superior purity. (In case of diesel engine, although its interior is contaminated particularly by combustion produced matter and deteriorated oil and thus such major troubles as friction increase and seizure of important parts and poor lubrication, if the oil's purity is good, various parts of engine are to be maintained clean. Therefore, extends the lifetime of engine greatly.)
8. Good defoaming property. (Prevents a drop in performance of pump due to air bubbles.)

2-2-4. Exchange of Engine Oil

1. Reasons for Exchange
   Engine oil is subjected to a high temperature during engine operation and, under such temperature, can be blended with air. Thus, the oil itself oxidizes and gradullay changes its property. Besides, water and impurities entered from external source, and fuel con-

taminate and dilute the oil, reducing the capability as a lubricating oil. As water and impurities are mixed into lubricating oil, there are emulsified compound and sludge which increase the viscosity of oil. In addition, if carbon particles deposited inside the cylinder enter the crankcase, oil therein becomes pitch black and can be seen to be worsened in quality at a look. If, however, deteriorated oil is continuously used, the reciprocating and/or rotating parts of engine interior are abrased or corroded, and finally possibility of seizure of bearings and cylinder beocmes great. For these reasons, used engine oil must be exchanged with new one at the time for which we describe below.

2. Exchange Time

Although the exchange time of engine oil differs from engine to engine, quality of lubricating oil, quality of fuel and engine operational condition, the first exchange time for DG grade oil should be following the total engine operation time of 20 hours from use of a brand new engine. The second time is at 30 hours following the first; the third and thereafter, 100 hours following the preceding exchange. The exchange should be conducted while the engine is still warm, discharging old oil completely and feeding new oil. For the exchange, however, observe the following cautions:

1) To exchange the oil inside the crankcase, remove old oil while the engine is warm, tilting the engine so that oil will rush out of the drain plug.
2) Prior to feeding new oil following the discharge of old oil, discharge soiled oil of the outlet side lub. oil strainer comptelely by turning the handle of the strainer several times, then by removing the screw at the lower part of the strainer, and finally by cranking the starting handle four or five times.
3) In course of overhaul of the engine, when the crankcase has been washed, wash oil is fully removed. Only then feed new engine oil. (Remember that for removing wash oil, do not use waste cloth because it might leave bits of down and dirt over the crankcase surface.

2-2-5. Cautions for Handling Engine Oils

At the time of exchanging, if different type of lubricating oil is to be used in place of old oil, pay your attention to the following cautions particularly: For an example, if DG gade heavy duty engine oil with purifier additive is to be fed to the engine using engine oil without additive, there will be no problem only if the engine is completely cleaned. However, if the sludge produced from the previously used oil is remaining inside the engine interior and through the pipe lines, this sludge is washed out due to purifying characteristic of the new oil. To prevent it is necessary to wash out engine interior as much as possible prior to feeding of the new oil. Adoption of the oil containing purifier additive for engine lubrication allows field oil to be soiled at much faster rate than at usual up to the second or third exchange time. It is thought not to be an evidence for poor oil but rather to clean the engine interior. If such is the case, therefore, it is actually a better sign for the engine.

To make the engine easier to start and ensure efficient distribution of the oil and best fuel economy it is essential to select the appropriate viscosity grade, which depends upon the ambient temperature.

For your guidance herebelow is given a table by which the most suitable lub. oil can be chosen easily.

Supplier	Brand Name	SAE No.			
		below 10 °C	10-20 °C	20-35 °C	over 35 °C
SHELL	Shell Rotella Oil	10W 20/20W	20/20W	30 40	50
	Shell Talona Oil	10W	20	30 40	50
	Shell Rimula Oil	20/20W	20/20W	30 40	
CALTEX	RPM Delo Marine Oil	10W	20	30 40	50
	RPM Delo Multi-Service Oil	20/20W 10W	20	30 40	50
MOBIL	Delvac Special	10W	20	30 40	
	Delvac 20W-40	20-40	20W-40		
	Delvac 1100 Series	10W 20-20W	20-20W	30 40	50
	Delvac 1200 Series	10W 20-20W	20-20W	30 40	50
ESSO	Estor HD	10W	20	30 40	
	Esso Lube HD		20	30 40	50
	Standard Diesel Oil	10W	20	30 40	50

— 14 —

3. PERIODICAL CHECKING & SERVICING

A periodical checking is necessary to keep the engine always in good condition. The frequency of the periodical checks may vary depending upon the purpose for which the engine is used, the conditions of use, the quality of oils used, and the method of handling of engine. It is, therefore, difficult to generalize on the frequency with which periodical checks and servicing should occur. However, here we will explain in general.

3-1. Periodical List of Items to be Checked and Frequency of Checks

Checking & Servicing Item		Service period					Illustration
		Daily	Every 50 hrs.	Every 250 hrs.	Every 500 hrs.	Every 1000 hrs.	
Fuel oil	(Prior to starting) Check fuel volume & replenish, if necessary	●					
	Discharge fuel tank drain to eliminate water condensation	●					
	Trun handle of fuel strainer	●					
	Discharge fuel strainer drain to eliminate particle build up		●				
	Clean fuel tank strainer				●		

— 15 —

Checking & Servicing Item		Service period					Illustration
		Daily	Every 50 hrs.	Every 250 hrs.	Every 500 hrs.	Every 1000 hrs.	
Lub. oil	Check lub. oil volume & replenish crankcase and reverse gear case, if necessary	●					
	Turn handle of lub. oil strainers	●					
	Discharge lub. oil strainer drains		●				
	Clean lub. oil strainers			●			
	Change lub. oil of crankcase			●			

Checking & Servicing Item		Service period					Illustration
		Daily	Every 50 hrs.	Every 250 hrs.	Every 500 hrs.	Every 1000 hrs.	
Lub. oil	Change lub. oil of crankcase			●			
	Change lub. oil of reversing gear case					●	
Cooling water	Check tightness of packing glands	●					
	Check cooling water circulation	●					
	Check anti-corrosive zinc				●		

Checking & Servicing Item		Service period					Illustration
		Daily	Every 50 hrs.	Every 250 hrs.	Every 500 hrs.	Every 1000 hrs.	
Fuel injection pump	Check oil feed to adjusting rod	●					
	Check atomization of fuel by priming	●					Good Bad
	Confirm fuel injection timing			●			
Fuel injection valve	Clean fuel injection valve strainer			●			
	Clean needle valve				●		
Cylinder head	Adjust valve clearance (suc. & exh. valves)			●			

— 18 —

Checking & Servicing Item		Service period					Illustration
		Daily	Every 50 hrs.	Every 250 hrs.	Every 500 hrs.	Every 1000 hrs.	
Cylinder head	Retighten cylinder head			●			
	Clean combustion chamber wall				●		
	Clean precombustion chamber				●		
	Lap the suc. & exh. valves				●		
	Check valve levers & valve guides				●		
Piston	Disassemble piston & check piston rings					●	

— 19 —

3-2. Checking Locations

In addition to all the periodical checks and servicing items, after 2500 hours of total engine operation, disassemble the engine completely, measure and inspect each part carefully and repair or replace all the excessively worn parts with new ones as required, with reference to "List of Wear Limit."

It is recommended that parts which will wear out before the next overhaul, should also be replaced even if they are within tolerance limits at this time. It is required, therefore, that a workshop equipped with facilities for measurement, inspection, and testing be available.

3-3. Checking Locations by Systems

1. Fuel System
   1) Clean fuel tank interior.
   2) Clean inside the fuel pipes with compressed air, and check for any crack in fuel pipes.
   3) Clean fuel strainer filter plates and interior.
   4) Check fuel injection condition and pressure of fuel injection system.
2. Lubricating Oil System
   1) Clean and inspect cylinder block interior and clutch housing.
   2) Clean and inspect lub. oil passages and lines inside the cylinder block, cylinder head and lub. oil cooler.
   3) Clean lub. oil strainer filter plates and interior.
   4) Check condition of lub. oil pressure regulating valve for strain and its spring tension.
3. Cooling System
   1) Check degree of deterioration of anti-corrosive zinc.
   2) Check extent of abrasion of cooling water pump valve and its seat. Also check the extent of abrasion and strain of cooling water pump valve spring.
   3) Check degree of abrasion of plunger reciprocating section.
4. Engine Body
   1) Check presence or absence of cracked cylinder head and looseness of insert.
   2) Inspect extent of abrasion of pre-combustion chamber.
   3) Check clearance between valve guide and valve shaft.
   4) Check the contact condition of the valve spring holder with valve lever and inspect the abrasion condition of contacting section of valve with lock pieces.
   5) Check strain and tension of valve spring.
   6) Check extent of abrasion of valve lever and valve lever bush.
   7) Check degree of contact of valve seat.
   8) Check abrasion of push rod and tappets.
   9) Check for scoring, excessive abrasion and bending of cam shaft.
   10) Check for scoring and excessive abrasion of and dimension measurement of each part of the cylinder, cylinder liner, crank journal metal, crankpin metal and crankshaft.
   11) Inspect for cracks in cylinder body.
       (Piston and Its Accessories)
   12) Check for cracks and abnormal contact of piston head and measurement of piston outer diameter.
   13) Check friction of piston top ring, trigger weld and cracking of piston ring land.

14) Check sliding and fitting conditions of piston's outer lateral surface.
15) Measure contacting state of outer circumference of piston and oil scraping rings, clearance between piston ring groove and piston ring as well as size of piston ring notch.
16) Measure clearance between piston pin and piston pin metal.

5. Crankshaft
   1) Measure clearance between a crankpin and crankpin metal.
   2) Measure clearance between crank journal and crank journal metal.

6. Miscellaneous
   1) Check for scoring and degree of contact of governor end bearing with governor 2nd lever.
   2) Check for scoring and degree of contact of governor weight with governor spindle.
   3) Check the contact surface condition of gear teeth and check for backlash.

4. MAINTENANCE STANDARDS OF MAIN PARTS

When an engine is used for a period of time, its parts will become worn, not only reducing the engines' performance, but also resulting in engine troubles unless these worn down parts are replaced. The lists of wear limits of main parts will be given at the end of this section. These wear limits are estimated values which will ensure superior performance of the engine and thus they are not absolute values by which the safety of worn down parts can be guaranteed. As indicated in the section concerning the periodical checks and servicing, it is recommended that the parts which will wear out before the next periodical check also be replaced, even if they are at the time of the check, not quite reaching their wear limits.

4-1. Maintenance Standards

(Unit in mm)

Part Name				Standard Dimension	Variation of Tolerance	Max. Allowable Clearance	Wear Limit	Correction Measure
Cylinder liner			Inner dia.	85φ	+0.030 / 0		When chrome plate is worn	Replacement of liner
			Lobe of rubber packing					Replacement when liner is extracted
Clearance of piston & cyl. liner	Inner dia. of piston		Top	85φ	-0.415 / -0.445		-0.30	Replacement of piston
			Skirt		-0.125 / -0.155			
	Inner dia. of cyl. liner				+0.030 / 0			
Commpression clearance				1.85	±0.2			
Piston ring	Clearance betw. piston ring and ring groove	Cut opening (VS. st'd dia.)	Comp. ring				1.5	Replacement
		Inside of liner	Oil scraper ring				1.5	Replacement
		Comp. ring	Ring breadth No. 1	3.5	-0.01 / -0.03	0.2	-0.10	Replacement of piston ring or piston
			Ring breadth No. 2, 3	2.5				
			Groove width No. 1	3.5	+0.025 / +0.010		+0.15	
			Groove width No. 2, 3	2.5				
		Oil scraper ring	Ring breadth	4.0	-0.01 / -0.03	0.2	-0.10	Replacement of piston ring or piston
			Groove width		+0.025 / +0.010		+0.15	
Piston pin			Boss inner dia. of piston	32φ	h_1 $\begin{smallmatrix}-0.004\\-0.017\end{smallmatrix}$	0.15	0.07	Replacement
			Outer dia. of piston pin		0 / -0.013			
			Bush inner dia. of piston pin		+0.060 / +0.030			
Crank shaft	Crank pin		Outer dia. of crank pin	64φ	-0.035 / -0.050	0.15	0.7	Re-polishing when the pin worn >0.06 and use of the underisize metal of either -0.25 or -0.5
			Bearing back metal inner dia.					
	Journal	Dia.	Outer dia. of journal	70φ	-0.040 / -0.055	0.15	0.7	
			Bearing dia.		+0.051 / +0.010			
		Breadth	Journal	219	+0 / +0.05	0.3	0.5	
			Standard bearing					

Part Name			Standard Dimension	Variation of Tolerance	Max. Allowable Clearance	Wear Limit	Correction Measure
Cam-shaft	Dia.	Journal section outer dia.	46φ	f_7 -0.025 / -0.050	0.15	-0.12	Replacement
		Bearing inner		H_2 +0.030 / 0			
	Breadth	Standard journal	35				
		Standard bearing	34				
	Height of cam		37.5	±0.01		-0.5	Replacement
Suction/exhaust valve	Clearance betw. valve push rod & valve guide	Suction valve	Valve push rod outer dia.	9φ	-0.040 / -0.055	0.25	-0.15	Replacement of valve or valve guide
			Valve guide inner dia.		H_2 +0.017 / 0			
		Exhaust valve	Valve push rod outer dia.	9φ	-0.040 / -0.055	0.20	-0.15	
			Valve guide inner dia.		H_2 +0.017 / 0			
	Valve seat	Angle		120°	+30 / 0			
		Sink depth		-0.7			1.6	Replacement of valve seat on exhaust side
	Clearance of head	Suction valve					0.10 ~ 0.20	
		Exhaust valve					0.10 ~ 0.20	

4-2. List of Measuring Positions

No.	Measuring Item	Measuring Position	Remarks	Measuring Instrument
1	Cylinder liner, inner diameter		"*" position in "a" & "b" directions	Cylinder gauge
2	Piston skirt, outer diameter		"*" position in "a" & "b" directions	Micrometer
3	Piston pin hole, inner diameter		"*" position in "a" & "b" directions	Cylinder gauge

No.	Measuring Item	Measuring Position	Remarks	Measuring Instrument
4	Breadth of piston ring grooves		"*" positions	Block gauge
5	Thickness of piston rings		"*" positions	Micrometer
6	Piston pin, outer diameter		"*" position in "a" & "b" directions	Micrometer
7	Piston pin metal, inner diameter		"*" position in "a" & "b" directions	Cylinder gauge
8	Crank journal, outer diameter and crank pin, outer diameter		"*" position in "a" & "b" directions	Micrometer
9	Crankpin metal, inner diameter		"*" position in "a" & "b" directions	Cylinder gauge
10	Crank journal metal, inner diameter		"*" position in "a" & "b" directions	Cylinder gauge
11	Clearance between crank journal and crank metal		"*" positions	Thickness gauge

No.	Measuring Item	Measuring Position	Remarks	Measuring Instrument
12	Suction/exhaust valve stem, outer diameter		"*" position in "a" & "b" directions	Micrometer
13	Suction/exhaust valve guide, inner diameter		"*" position in "a" & "b" directions	Cylinder gauge
14	Cylinder head valve seat, settle depth		"*" positions	Depth gauge
15	Clearance between piston pin and piston pin metal		Calculate from measurements 6 & 7	
16	Clearance between crank pin and crank pin metal		Calculate from measurements 8 & 9	
17	Clearance between crank journal and crank journal metal		Calculate from measurements 8 & 10	
18	Clearance between suction/exhaust valve and suction/exhaust valve guide		Calculate from measurements 12 & 13	

4-3. List of Wear Limit

(Unit: mm)

		Standard Dimension	Wear Limit	Remarks
Dia. of cylinder liner (inner)		85φ	+0.13	
Dia. of piston skirt (outer)		85φ	−0.30	
Dia. of piston pin hole (inner)		32φ	+0.03	
Width of piston ring groove	Compression ring	$\frac{3.5}{2.5}$	+0.15	
	Oil scraping ring	4.0	+0.15	
Width of piston ring	Compression ring	$\frac{3.5}{2.5}$	−0.10	
	Oil scraping ring	4.0	−0.10	
Dia. of piston pin (outer)		32φ	−0.07	
Dia. of piston pin metal (inner)		32φ	Max. clearance between pin & metal is 0.15	
Dia. of crank journal (outer)		70φ	−0.70	
Dia. of crankpin (outer)		64φ	−0.70	
Dia. of crank journal metal (inner)		70φ	Max. clearance between journal and metal is 0.15	It is possible to be provided with under size metal.
Dia. of crankpin metal (inner)		64φ	Max. clearance between metal and pin is 0.15	It is possible to be provided with under size metal.
Side clearance between crank journal and its metal		0.15 ~ 0.3		It is possible to be provided with over size metal. For regulating decrease or increase the packings of metal body.
Dia. of suction & exhaust valve rods (outer)		9φ	−0.15	
Dia. of suction & exhaust valve guides (inner)	Suction side	9φ	Max. clearance between valve rod and guide is 0.25	
	Exhaust side	9φ	Max. clearance between valve rod and guide is 0.20	
Depth of cylinder head valve seat	Suction side	0.7	+2.0	
	Exhaust side	0.7	+2.0	

4-4. List of Undersize Metals

The following undersize metals are also available as optional parts.

(Unit: mm)

Part Name		Standard Crankshaft (Dia. x Width)	Code No.	Undersize Crankshaft Dia.	Symbol Marked	Remarks
Crank pin metal		64φ x 33	124210–23800	0.25	UO 25	Kelmet metal
			124210–23810	0.50	UO 50	
Crank journal metal	Flywheel side	70φ x 50	124210–02120	0.25	UO 25	Kelmet metal
			124210–02140	0.50	UO 50	
	Gear side	70φ x 50	124210–02130	0.25	UO 25	Kelmet metal
			124210–02150	0.50	UO 50	

4-5. List of Oversize Metals

The following oversize metals are also available as optional parts.

(Unit: mm)

Part Name		Standard Crankshaft (Dia. x Width)	Code No.	Oversize Metal Width	Symbol Marked	Remarks
Crank journal metal	Flywheel side	70φ x 50	124210–02120	0.10	UO 25	Kelmet metal
			124210–02140	0.20	UO 50	
	Gear side	70φ x 50	124210–02130	0.10	UO 25	Kelmet metal
			124210–02150	0.20	UO 50	

5. ENGINE DISASSEMBLY

5-1. Precautions Prior to the Disassembly of the Engine

Prior to the disassembly of the engine the following precautions should be observed:
1. A clean, dust-free workshop should be available for the disassembly.
2. Prepare a table or board on which the disassembled parts can be placed and covered to prevent them from damage or loss before reassembly.
3. Washing oil and a tin can are required for cleaning during engine disassembling.
4. Prepare the following tools.
    - Standard accessory disassembling tools and general tools:
      1) Spanners (10×14, 17×19, 21×23 & 26×32)
      2) Special spanner (10×14)
      3) Special spanner (23)
         (employed with handle of box spanner of 8)
      4) End-nut spanner
      5) Monkey wrench
      6) Screwdriver, minus
      7) Screwdriver, plus
      8) Box spanner (21) with handle
      9) Nozzle strainer removing tool
      10) Delivery valve guide removing tool
      11) Lapping powder, canned
      12) Valve lapping tool
      13) Thickness gauge, 0.2mm
      14) Suction/exhaust valve removing & inserting tool
      15) Fuel injection time adjusting gauge
      16) Fuel injection time adjusting spanner (17)
      17) Special box spanner (46)
      18) Pliers
      19) Hammer
      20) Hammers (Lead, copper & plastics)
      21) Eyenut
      22) Circlip pliers
      23) Pin set
      24) Box wrench
      25) Eyenut wrench

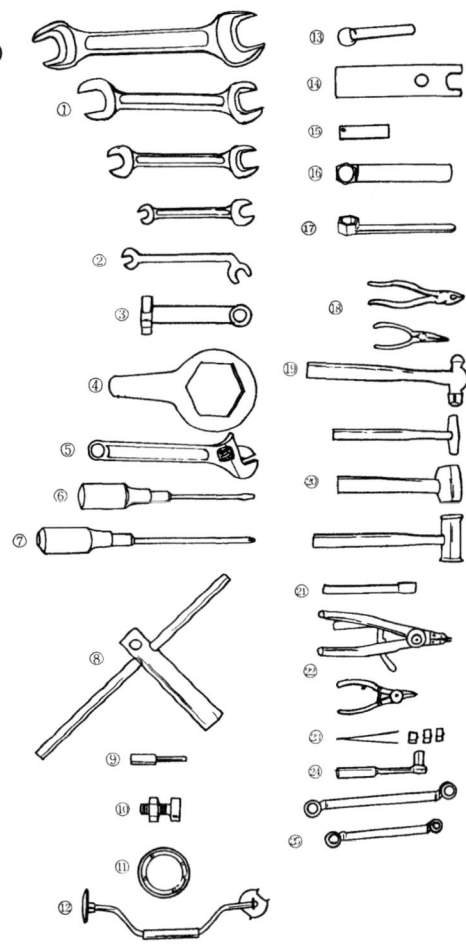

— 28 —

- Special tools for repairs:
1) Gear removing tool, complete
2) Piston inserting tool
3) Cylinder liner removing tools, complete
4) Freewheel removing tool, complete
5) Suction/exhaust valve seat cutter, complete

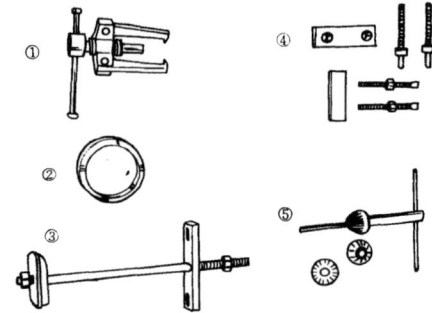

- Measuring instruments:
1) Vernier calipers
2) Micrometer
3) Cylinder gauge
4) Thickness gauge
5) Depth gauge
6) Tachometer
7) Torque wrench
8) Fuel injection valve tester
9) Fuel injection pump tester

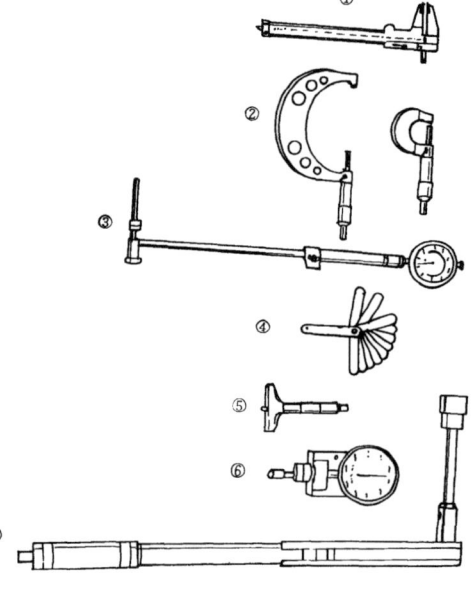

- Other items to be prepared:
1) White paint
2) Sealing agent
3) Red lead
4) Fine lapping powder
5) Chromium oxide

5-2. General Cautions for Maintenance and Cleaning

The following cautions should be observed for maintenance and cleaning:

1. Make a good visual inspection of each part for carbon deposits and scoring. This practice will be very helpful for maintenance.
2. Be careful not to damage the parts when removing the carbon deposits.
3. Avoid using a wire brush or sandpaper for cleaning the precision parts, such as the contact surface of suction and exhaust valves, the reciprocating surface of plungers, etc.
4. Use only clean washing oil for final washing.
5. Thoroughly clean the outside of engine block and the inside of crankcase at the time of disassembly. Prior to reassembly also clean those parts and places which are difficult to clean afterward.
6. Clean the oil holes subject to lub. oil sludge setting and the inner surface of lub. oil pipe not only with washing oil but also pressurized air thoroughly.
7. Clean any rusty part with fine sandpaper and apply oil afterward.
8. If any part is found scored, replace that part.
9. If a mark or small dent is made by striking the part accidentally, remove such mark or dent by rubbing with an oil-stone.
10. After fitting new parts, carefully inspect them for the state of fitting, contact, and clearance.

5-3. Cautions for Disassembling the Engine

The following cautions should be observed for disassembling the engine:

1. Disassemble the engine according to given disassembly procedure.
2. Do not disassemble any part unnecessarily.
3. Do not scratch or damage any part.
4. Use only proper tools.
5. Provide appropriate locks when disassembly work is difficult as such parts as shafts tend to turn with their accessories.

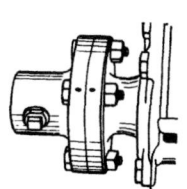

Drive shaft & Propeller shaft

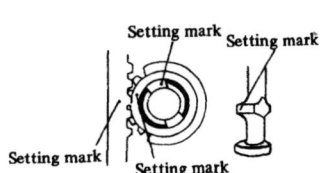

Pinion & Rack

6. For removing of such parts as shafts, use appropriate brass or copper rod and caulking plates.
7. Arrange the disassembled parts in good order.
8. Confirm the positions by setting marks.
   The following parts are provided with the setting marks:
   Connecting rod large end and its metal bush; connecting rod metal bush and rod bolt; camshaft gear and crank & gear; clutch housings A, B & C; shaft couplings of thrust shaft and propeller or intermediate shaft; fuel oil control pinion and fuel oil adjusting rack of fuel injection pump; and fuel oil control pinion and plunger of fuel injection pump.
9. Make the most out of disassembly by cleaning those sections usually difficult to clean.
10. Wash clean the disassembled parts and arrange in order.

5-4. Procedure on Disassembling

Part Name	Disassembling Procedure	Tool Used	Remarks	Illustration
Governor	Remove the regulator spring.	By hand		
	Remove the pin of governor second lever (with a split pin).	Pliers		
	Remove the governor cover.	Spanner (14)	With rubber washer	
	Remove the governor.	By hand	With paper packing	
	Remove the cover of reversing gear upper case.	Spanner (19)	With paper packing	
	Set the clutch handle to neutral position.	By hand		
Remote control device	Remove the rear bearing cover of ahead shaft.	Spanner (17)	With paper packing	
	Remove the operating lever together with the clutch handle.	By hand		

Part Name	Disassembling Procedure	Tool Used	Remarks	Illustration
Remote control device	Remove the sliding Shaft Pin (with split pin).	Pliers		
	Remove the rear bearing box of ahead shaft.	Spanner (17) Copper hammer	With paper packing	
Clutch upper side case	Take out the oil gauge rod of reversing gear.	By hand		
	Remove the upper case of reversing gear (with a knock).	Spanner (17) Spanner (14)	With paper packing 3 pcs. of case fitting bolts have packings.	
Clutch body	Remove the clutch body fitting bolts (with bend washer).	Hammer Screwdriver Spanner (10)		

Part Name	Disassembling Procedure	Tool Used	Remarks	Illustration
Clutch body				
	Fit the hanging bolt to the screw hole of clutch housing Ⓑ and remove the clutch body.	Hanging bolt By hand		
	Remove the fitting bolts of clutch housings Ⓐ Ⓑ Ⓒ (with bend washers).	Hammer Chisel Spanner (14)		
	Remove the clutch housing Ⓐ and then the sliding shaft.	Copper hammer		

Part Name	Disassembling Procedure	Tool Used	Remarks	Illustration
Clutch body				
	Remove the clutch housing ⓑ.	Copper hammer		
	Remove V-lever.	By hand		
	The front bearing of ahead shaft can be taken off. Remove the clutch housing ⓒ by pushing it to the right.	By hand	As the neutral cut may protrude it is necessary to remove it gradually.	
	Remove the friction disc fitting bolts of ahead shaft (with bend washers) and then take out the friction disc and the doubling plate.	Hammer Chisel Spanner (14)		

Part Name	Disassembling Procedure	Tool Used	Remarks	Illustration
Clutch body				
	Remove the friction disc retainer.	By hand		
	Remove the shake-proof nut of ahead shaft rear bearing (with bend washer).	Hammer Screwdriver Special box spanner (46)		
	Remove the rear roller bearing of ahead shaft together with the reduction gear by means of gear removing tool.	Gear removing tool		
	Remove the key.	Hammer Screwdriver		
	Remove both ahead and astern shafts.			

Part Name	Disassembling Procedure	Tool Used	Remarks	Illustration
Cluch body	In this case the front and rear ball bearing as well as the springs of astern shaft can be dismounted.			
	After taking off the fitting bolts of astern shaft friction disc (with bend washers), remove the friction disc and caulking plate.			
Intermediate gear shaft	Remove the fitting nuts of intermediate gear shaft (with a split pin).	Pliers Spanner (32)		

Part Name	Disassembling Procedure	Tool Used	Remarks	Illustration
Intermediate gear shaft	Tap the rear part of intermediate gear shaft with a copper hammer and remove the shaft together with the intermediate gear.	Copper hammer		
Thrust shaft	Remove the rear cover of thrust shaft (with an oil seal).	Spanner (17)	With paper packing	
	Remove the front shake-proof nut of thrust shaft (with bend washer).	Hammer Screwdriver Special spanner (46)		
	Tap the front part of thrust shaft with a lead hammer and take out the thrust shaft together with the rear ball bearing of thrust shaft, the washer and the regulating plate to the back side. The large gear and the front ball bearing of thrust shaft are left in the case when they are removed. Take away the under case of reversing gear.	Lead hammer Spanner (17)		

Part Name	Disassembling Procedure	Tool Used	Remarks	Illustration
Cylinder head assembly	Remove the fuel leak pipe (injection valve). Remove the fuel high pressure pipe. Remove the injection valve. Remove the chain cover. Remove the chain. Remove the stay of the chain starting handle. Take out the bonnet.	Spanner (17) Spanner (19) Spanner (17) Spanner (14) Minus screw-driver Spanner (21) By hand		
	Remove the stay of the valve arm shaft.	Spanner (19)		
	Remove the push rod. Remove the lubricating oil of valve arm chamber.	By hand Spanner (17)		
Cylinder head	Remove the cylinder head.	Box spanner (23)		

Part Name	Disassembling Procedure	Tool Used	Remarks	Illustration
Cylinder head				
Protective cylinder of Push rod cover	Remove the push rod cover.	By hand		
Gasket packing	Remove the gasket packing.	By hand	With O-ring	
Side plate of crankcase	Remove the oil gauge rod of crankcase. Remove the side plate of crankcase.	By hand Spanner (14)		
Connecting rod bolt	Remove the rod bolt (with bend washer).	Spanner (19)		
	Remove the metal holder.	By hand		

— 39 —

Part Name	Disassembling Procedure	Tool Used	Remarks	Illustration
Piston	After pushing the piston conrod upward withdraw it.	By hand		
Fuel pump	Remove the fuel pipe (strainer ~ pump). Remove the connecting pin of fuel pump (with split pin).	Spanner (21) Pliers		
	Remove the fuel pump.	Spanner (19)		
Roller guide	After taking out the lock bolt, withdraw the roller guide.	Spanner (14) By hand		
Cooling water pump	Remove the cooling water tube (pump cooler).	Spanner (14)		

Part Name	Disassembling Procedure	Tool Used	Remarks	Illustration
Cooling water pump	Remove the cooling water pump.	Spanner (19)		
	Remove the plunger pin of cooling pump with bend washer.	Hammer Screwdriver Spanner (19)		
	Withdraw the plunger.	By hand		
Gear box	Remove the clamping shake-proof nut of crank gear (with bend washer).	Special box spanner (46)		

Part Name	Disassembling Procedure	Tool Used	Remarks	Illustration
Gear box	Remove the crank gear by means of a gear drawing-out tool.	Gear drawing-out tool		
	Remove the lubricating oil pipe (pump ~ strainer).	Spanner (23)		
	Remove the lubricating oil pump.	Box wrench (14)		
Flywheel	Remove the flywheel clamping nuts (with bend washers).	End-nut spanner		

Part Name	Disassembling Procedure	Tool Used	Remarks	Illustration
Flywheel	Withdraw the flywheel by means of drawing-out tool.	Flywheel drawing-out tool		
	Remove the flywheel	By hand		
	Remove the flywheel key.	Screwdriver		
Camshaft	Remove the flywheel clamping nuts (with bend washers).	Spanner (32)		
	Withdraw the flywheel and chain wheel together by means of drawing-out tool.	Flywheel drawing-out tool		

Part Name	Disassembling Procedure	Tool Used	Remarks	Illustration
Camshaft	Remove the camshaft front cover.	Plus screw-driver		
	Remove the cam gear side plate.	Spanner (14)		
	Remove the cam gear clamping bolts (with bend washers).	Spanner (32)		
	Remove the cam gear from the camshaft by means of drawing-out tool.	Drawing-out tool	Be careful so that there is no interference between the tappet and the cam.	
	Withdraw the camshaft forward. At that time the cam gear, the pump connecting rod and the tappet can be removed.	By hand		

Part Name	Disassembling Procedure	Tool Used	Remarks	Illustration
Camshaft	After taking out the setting bolts of camshaft front metal, withdraw the metal.	Spanner (14) Hammer Caulking plate		
	Remove the fitting bolts of retaining plate for camshaft rear ball bearing (with bend washers) and take off the retaining plate and the ball bearing.	Spanner (14) Hammer Caulking plate		
	Remove the lubricating oil pipe (oil pressure regulating valve cooler ~ metal body).	Spanner (19) Spanner (21)		
Crankshaft	Remove the crankshaft front cover (with oil seal).	Plus screwdriver	With paper packing	
	Remove the metal body fitting nuts.	Spanner (17)		

— 45 —

Part Name	Disassembling Procedure	Tool Used	Remarks	Illustration
Crankshaft	Screw the bolt for withdrawing the metal body into metal body drawing-out hole and then take out the metal body.	Metal body drawing-out bolt	With paper packing	
	Remove the crankshaft.	By hand	In case removing the crankshaft, the balance weight of No. 1 cylinder should be placed on the part of cylinder block.	
Lubricating oil cooler	Remove the lubricating oil pipe (cooler-oil pressure gauge).	Spanner (17)		
	Remove the lubricating oil cooler.	Spanner (14)	With paper packing	

— 46 —

6. ENGINE REASSEMBLY

6-1. Cautions for Reassembling the Engine

The following cautions should be observed for reassembling the engine:

1. Be sure that the disassembled parts have been thoroughly washed.
2. Carry out the reassembly according to the reassembling procedure.
3. Use only the proper tools.
4. Apply lub. oil to such sections as rotating and interlocking sections.
5. Do not scratch such parts as "O" rings, oil seals and packings.
6. Apply proper locks if tightening of such parts as the shaft are difficult because they tend to turn together with their accessories.
7. For tapping-in such parts as shafts apply appropriate caulking plates and brass or copper rod.
8. Reassemble first the section which was disassembled in the detail.
9. Check on abrasion and scratched conditions of various parts, and if necessary replace them with the new ones. Repairable scratches and scars should be corrected.
10. Tighten bolts and nuts uniformly little by little.
11. Do not use old split pins, bend washers and packings.
12. Use split pins and wires that fit their set holes perfectly.
13. After fitting the split pins and bend washers, be sure to split the pins and to bend the tips of washers.
14. Match the setting marks and cuts and center the setting marked parts.
15. After fitting the shafts and rotating section, check if they turn smoothly, and proceed to next reassembly step.
16. Check for proper clearances before proceeding to the next reassembly step.
17. After completion of reassembly, adjust the parts which call for such adjustment.

6-2. Procedure on Reassembling

Part Name	Reassembling Procedure	Tool Used	Remarks	Illustration
Lubricating oil cooler	Fit the lubricating oil cooler.	Spanner (14)	Do sufficient packing	
	Attach the lubricating oil pipe (oil pressure gauge of the cooler).	Spanner (17)	With paper packing	

Part Name	Reassembling Procedure	Tool Used	Remarks	Illustration
Crankshaft	Insert the crankshaft	By hand	(Insert No. 1 cylinder side balance weight through the cylinder block).	
	Fit the metal housing.	Spanner (17)	(With paper packing). Adjust the side clearance of crankshaft to be 0.15 ~ 0.3mm increasing or decreasing the metal housing packing.	
	Attach the fore-lid of crankshaft (with oil wool).	Plus screw-driver	(With paper packing). Be sure to fit the lid with the oil escape hole below.	
	Insert the lubricating oil pipe (metal body ~ cooler ~ oil pressure adjusting valve).	Spanner (19) Spanner (21)		
Camshaft	Insert the camshaft rear side ball bearing into the cylinder block and fit the bearing holding plate with bolts (with bend washers). Bend the bend washers.	Hammer Caulking metal Spanner (14) Hammer Driver		

— 48 —

Part Name	Reassembling Procedure	Tool Used	Remarks	Illustration
Camshaft				
	Insert the camshaft front side metal into the cylinder block and set with bolts.	Hammer Caulking metal Spanner (14)	Set the metal oil hole to the inside of the upper side cylinder block. (The oil escape groove is set on the lower side) and match the metal setting holes.	
	Insert the tappet into the tappet hole of cylinder block, and insert the spacer and cam gear (with cooling water pump connecting rod) inserting the cam shaft from fore side.	Lead hammer Caulking metal	Insert the spacer so that its wider side comes to cam gear side. Oil hole of cooling water pump connecting bar should be up.	
	Lock the camshaft and cam gear with cam gear bolt (with bend washer) and bend the bend washer.	Spanner (32) Hammer Caulking metal		

Part Name	Reassembling Procedure	Tool Used	Remarks	Illustration
Camshaft	Fit the cam gear side lid. Fit the cam gear front lid.	Spanner (14) Plus screw-driver	(With paper packing) (With paper packing)	
	Insert the free wheel and chain wheel together into the camshaft and lock it with bolts (with bend washer), and bend the bend washer.	Copper hammer Spanner (32)	(Set the free wheel so that its removing threaded hole comes to the front side.)	
Flywheel	Insert the key with fly-wheel, and lock it with nuts (with bend washer) and bend the bend washer.	Hammer End-nut spanner Caulking metal		

— 50 —

Part Name	Reassembling Procedure	Tool Used	Remarks	Illustration
Flywheel				
Gear box	Fit the lubricating oil pump and crank gear after checking the back-lash of lubricating oil pump driving gear.	Box spanner (14)	(Back-lash 0.08 ~ 0.16mm)	
	Attach the lubricating oil pump (pump ~ strainer).	Spanner (23)		
	Insert the crank gear together with baffle plate matching the matching mark with cam gear, and fasten the crown nut (with bend washer) and bend the bend washer.	Special box spanner (46)	The baffle plate should be put in the front side of crank gear. Be sure to match the matching marks of crank gear and cam gear.	

Part Name	Reassembling Procedure	Tool Used	Remarks	Illustration
Gear box			Fasten sufficiently with crown nut preventing the bend washer of crown nut from being caught by the screw thread of crankshaft.	
Cooling water pump	Insert the cooling water pump plunger and set it with cooling water pump connecting rod. Bend the bend washer.	Spanner (19) Hammer Screwdriver		

Part Name	Reassembling Procedure	Tool Used	Remarks	Illustration
Cooling water pump	Install the cooling water pump.	Spanner (19)		
	Attach the cooling water pipe (pump ~ cooler).	Spanner (14)		
Roller guide	Attach the roller guide and locking bolt.	Spanner (14)		
Fuel pump	Install the fuel pump.	Spanner (14)		
	Fit the coupling pin (with split pin) of fuel pump and couple the pump. Install the fuel oil pipe (pump ~ strainer).	Pliers Spanner (23)		

Part Name	Reassembling Procedure	Tool Used	Remarks	Illustration
Piston	After setting the crankpin part around the top position, insert piston from the upper side of cylinder liner using piston inserting tool.	Piston inserting tool.	Do not line up the piston ring cuts when inserting piston. Set the crankshaft end of connecting rod to face it to non-operating side.	
Connecting rod	Insert metal holder and clamp with rod bolt (with bend washers). Bend the bend washers.	Spanner (19)	Set the matching numbers of metal holder and connecting rod. (Set the matching numbers of rod bolt and metal holder.) (Rod bolt locking torque: 6 kg/m)	

Part Name	Reassembling Procedure	Tool Used	Remarks	Illustration
Crankcase side lid	Fit the crankcase side lid. Attach the crankcase side oil level gauge.	Spanner (14) By hand	(With paper packing)	
Gasket packing	Attach the gasket packing of cylinder head.		Be sure to match the cooling water oil seal.	
Push rod cover	Attach the push rod cover.		Do not forget to set O-rings at the upper and lower fitting parts.	
Cylinder head	Insert the cylinder head.	By hand	Set the cylinder head fitting part (with O-rings) of push rod. Keep O-rings from being caught.	
	Fasten the cylinder head.	Torque wrench	Clamp it tucking. (Locking torque: 15 kg/m)	

Part Name	Reassembling Procedure	Tool Used	Remarks	Illustration
Cylinder head assembly	Fit the push rod. Place the valve lever shaft support. Set the valve lever case lubricating oil pipe. Attach the fuel jet pump. Attach the high pressure fuel pipe. Attach the bonnet. Set the chain starting handle shaft support. Set the chain. Attach the chain cover. Fit the fuel leak pipe.	By hand Spanner (19) Spanner (17) Spanner (17) Spanner (19) Spanner (21) Minus screw driver Spanner (14) Spanner (17)		
Clutch lower case	Fix the paper packings to the jointed faces of the upper and lower cylinder block reversing gear cases using adhesives and thus the lower reversing gear case assembly is fitted.	Spanner (17)	(Paper packing)	
Thrust shaft	Put the thrust shaft rear lid (with oil seal) to the thrusting shaft and its packing to the thrust shaft rear part ball bearing, spacer, adjusting liner, and master reduction gear key, and put the master reduction gear and thrust shaft front ball bearing.	Special box spanner (46)		

— 56 —

Part Name	Reassembling Procedure	Tool Used	Remarks	Illustration
Thrust shaft	Attach the thrust shaft rear lid.	Spanner (17)		
Intermediate gear shaft	Put the intermediate gear and intermediate gear shaft and set it into the case by means of knocking the shaft front with hammer.	Lead hammer	Be sure to match the shaft oil holes.	
	Insert washers and clamp it with nuts and open it by split pin.	Spanner (32)		
Clutch assembly	Attach the friction plate to the astern shaft inserting washers and clamp it with bolts (bend washers) and bend the bend washer. Set the astern shaft side ball bearing.			

Part Name	Reassembling Procedure	Tool Used	Remarks	Illustration
Clutch assembly				
	Put a friction plate holder between the friction plates of astern shaft and ahead shaft sides, and combine the astern shaft and ahead shaft.			
	Put a spring into the astern shaft, and set the rear ball bearing.			
	Attach the reduction pinion key.			
	Attach the reduction pinion.			
	Clamp the clutch housing (assembly) with bolts (with bend washers) matching the mark with crank gear. Bend the bend washers.	Spanner (17) Hammer Screwdriver		

— 58 —

Part Name	Reassembling Procedure	Tool Used	Remarks	Illustration
Clutch assembly				
Clutch upper case	Put the paper packing using adhesive to the joint face of the upper and lower cases of reversing gear to set the upper case of reversing gear. Insert a knock and fasten it with bolt.	Hammer Spanner (17)	Use copper packing washer for the case clamp bolt in the left figure.	

Part Name	Reassembling Procedure	Tool Used	Remarks	Illustration
Clutch upper case				
Remote control device	Attach the oil gauge rod of reversing gear. Set the neutral notch hole side of sliding shaft to the fuel oil side. Fit the rear bearing box of ahead shaft. Set the notch to the neutral position.	By hand By hand Copper hammer Spanner (10)	The sliding shaft should be at the neutral position. Remove the neutral notch and spring which should be set after bearing box has been fixed. (With paper packing).	
	Attach the sliding shaft pin (with split pin).	Pliers		
	Fit the operating lever fork (with clutch hand).	By hand		
	Fit the rear bearing box cover of ahead shaft. (Check clutch operation.)	Spanner (17)	(With paper packing)	

Part Name	Reassembling Procedure	Tool Used	Remarks	Illustration
Governor	Attach the governor.	By hand	(With paper packing). The oil escape hole should be lower.	
	Fit the governor cover.	Spanner (14)	(With paper packing). Be sure to use washer with rubber for the fitting bolt.	
	Couple the governor No. 2 lever and governor ring with a pin (with split pin).	Pliers		
	Couple the governor No. 2 lever and speed control handle with a regulator spring.	By hand		

Part Name	Reassembling Procedure	Tool Used	Remarks	Illustration
Clutch	Fit the roller bearing mounted with spacer, and fasten it with crown nuts (with bend washers). Bend the bend washers.			
	Fasten the friction plates on the ahead shaft inserting washers with bolts (with bend washers), and bend the bend washers.			

Part Name	Reassembling Procedure	Tool Used	Remarks	Illustration
Clutch	Set a lever for the V lever on the friction plate using pins.			
	Put the neutral cut (with spring) and insert it from the rear side of housing ⓒ.		Be sure to match the neutral cut of friction plate and the notch hole of housing ⓒ.	
	Insert the housing Ⓑ (with ahead shaft front ball bearing) into the housing ⓒ (ball bearing to ahead shaft).		Match the matching No. of housing ⓒ and Ⓑ.	
	Put the sliding shaft from the front side and insert the roller pin (with spring) into the sliding shaft.			

Part Name	Reassembling Procedure	Tool Used	Remarks	Illustration
Clutch	Fit V lever to the lever with a pin and match it with the roller pin.			
	Fasten the housing Ⓐ to Ⓑ with bolts (with bend washers) and bend the bend washers.		Be sure to match the matching No. of housing Ⓐ and Ⓑ.	

7. DISASSEMBLY & REASSEMBLY OF OTHER PARTS

7-1. Fuel Injection Pump

If fuel oil contains dusts, water, and other impurities or is of low quality, it will cause abnormal abrasion of plunger and delivery valve, making it impossible to send the right amount of fuel delivery from the fuel injection pump. A similar malfunction occurs when the contact surface of the delivery valve seat is improper or when dust is found on these.

1. Inspection
   - Inspection Method 1
     1) Attach a nozzle tester to the fuel injection pump in its fitting state, fix the ready adjusted fuel injection valve to the fuel high-pressure pipe so that atomization of fuel may be checked when fuel is to be injected from the valve.
     2) Prime the fuel injection pump, and check on the atomization of fuel at the fuel injection valve.
     3) If atomization of fuel is found to be inferior, this means that there is a malfunction in a part of the fuel injection pump.

   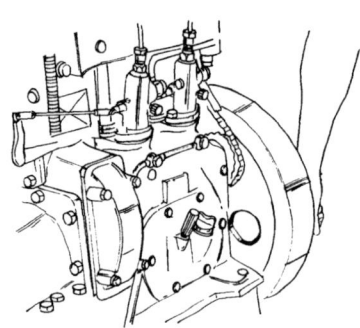

   - Inspection Method 2
     (Employing fuel injection pump tester)
     1) Fix the fuel injection pump to the pump tester, and deliver fuel to the pump after the adjusting rack is set at the increased injection volume side. Then measure the delivery pressure resistance.
     2) If the reading of the pressure gauge indicates above 500 kg/cm^2, the plunger of the pump is working normally.
     3) Next, discontinue the delivery action of the pump and check on the pressure withstanding time, observing the pressure gauge. If this pressure withstanding time is extremely short, this means that there is a malfunction in the delivery valve of the pump. Thus, if any part of the pump is found to be defective, the following disassembly and repairing become necessary.

   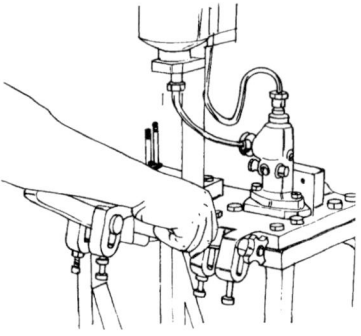

2. Disassembly
   1) Remove the delivery valve spring holder to take out both the delivery valve spring and the delivery valve itself.
   2) Use the delivery valve guide removing tool (standard accessory) when the delivery valve guide is to be taken out. Continue the removal by putting the washer on the upper surface of the pump body, screwing in the delivery valve guide removing bolt to the threaded part of the guide along with the removing bolt nut, and tightening this nut.
   3) When the plunger guide circlip is taken out, the plunger guide, plunger spring retainer, plunger spring, plunger spring holder and fuel control ring can be pulled down.

4) Remove the plunger barrel lock with copper packing, apply a brass or copper rod to the lower part of plunger barrel, and then hammer the barrel lightly as it is pushed up along with the copper packing.

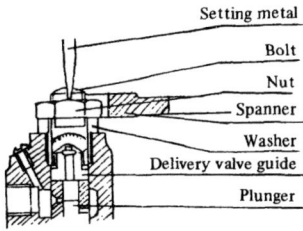

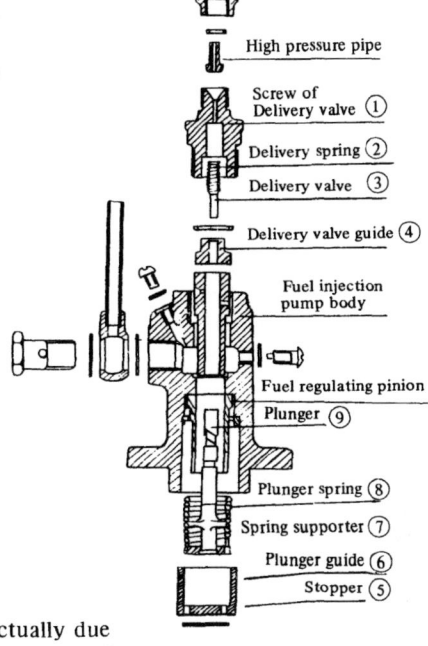

3. Service

   Malfunctions of the fuel injection pump are actually due to 1) a defective delivery valve and/or 2) a malfunction of the plunger and plunger barrel.

   1) If there is poor contact between the delivery valve and the valve seat of the valve guide, apply lapping powder to the valve seat and lap it. For this lapping, use only the fine lapping powder, and finish up the lapping using the chromium oxide. However, if the scar on the valve seat is a deep one and lapping of such seat would not correct the situation or if the valve seat has been worn excessively, replace the valve seat with a new one. Replace it along with a new delivery valve and valve guide.

   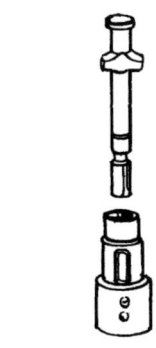

   2) Insert the plunger into the plunger barrel, and check on the clearance for the degree of abrasion. If there is a considerable clearance or if the working tip part of plunger is without luster and becomes white, replace the plunger with a new one along with a new plunger barrel. Handle the plunger and barrel carefully because they are precision machined and lapped together as a single part.

4. Reassembly

   1) It is very important to pay attention to handling the various parts of the fuel injection pump as they are precision machined and may not be scratched or scarred. Prior to reassembling them, be sure to wash them well in flushing oil, and particularly do not allow any dirt to enter the plunger and plunger barrel. It is recommended that lub. oil be applied to these parts for rust preventing purposes.

2) Reassemble the pump in the opposite order of disassembly.
3) Pay special attention to copper packings for attaining oil tightness.
   a. Copper packing between the plunger barrel and the pump body.
   b. Commpper packing of plunger barrel lock screw (Elimination of this packing causes introduc tion of air into the pump interior).
   c. Copper packing between the plunger barrel and the delivery valve guide.
4) There are setting marks at the following positions; be sure to match them.
   a. Rack of adjusting rod and pinion of fuel control ring.
   b. Fuel control ring and plunger.
      Note: *This injection pump has been designed so that if its adjusting rod moves to the scaled side, the injection volume increases and consequently the injection time advances accordingly.*
5) Attach the completely assembled fuel injection pump to the rest of the engine according to the procedure on engine reassembly.

7-2. Adjustment of Governor

A governor system, fuel injection volume control equipment, is very sensitive to load variation and functions to control the fuel injection pump. In other words, the governor corresponds to the nervous system of the human body. If the governor does not work correctly, the required fuel injection volume and consequently the desired brake horsepower will not be obtained.

1) Lower the governor handle, and confirm that the fuel control rack easily shifts to right and left.

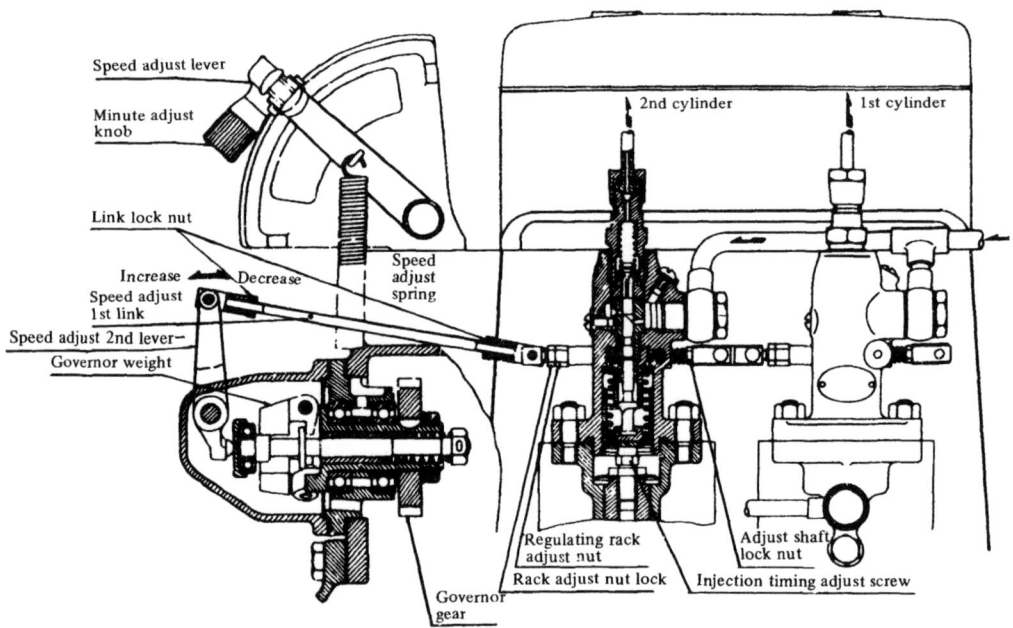

2) Tighten the lock nut on the adjusting nut of the adjusting rack for each cylinder so that the same amounts of adjusting nut of adjusting rack and adjusting lever of adjusting shaft will be retained for all of the cylinders.
3) Adjust the adjusting shaft so that a two-point mark, engraved on the adjusting rack of each cylinder's fuel injection pump, matches with the A surface of the fuel injection pump for all such shafts, taking any one of the cylinders as a base, and then tighten the lock nut. In other words, adjust the connecting section of the governor so that the two-point mark engraved on the adjusting rack of each cylinder's injection pump comes to the A surface of each injection pump at the same time.

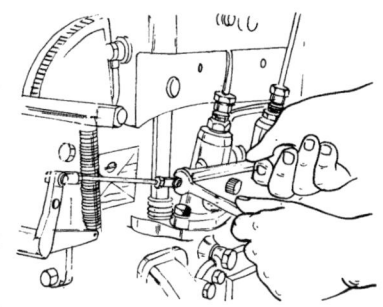

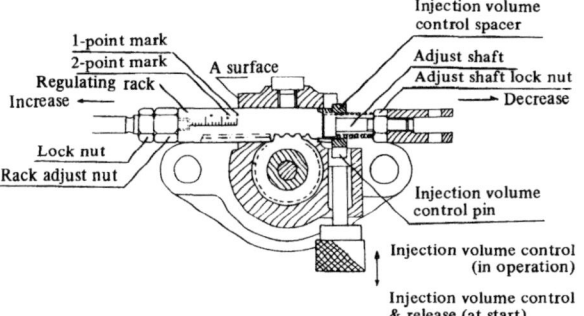

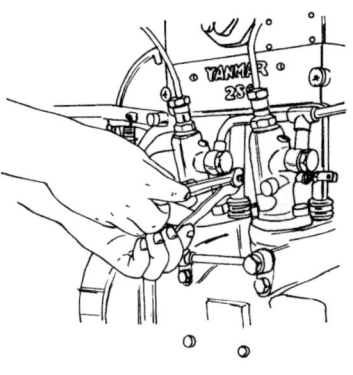

4) Set the governor handle to the full speed position.
5) Loose lock nuts from the ends of governor link.
6) Pull the fuel injection limiting pin from the fuel injection pump of each cylinder, and turn the governor link until the left end of adjusting rack comes in contact with the adjusting nut, provided that the adjusting rack is at the extreme left (or is at the matching point of its one-point mark with the A surface of the injection pump body). At this position, tighten up the lock nuts of the governor link. That is, when the governor handle is set at the full speed position, without pulling the limiting pin of the injection pump, push, by hand, the governor link to the right (or in the direction of increasing fuel injec-

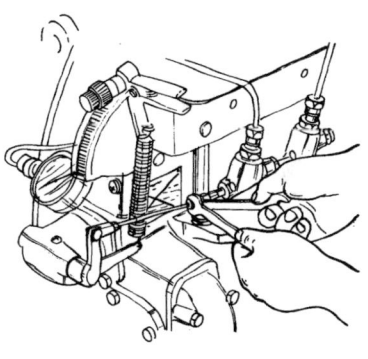

tion). This will give a clearance of about 1mm between the left end of the adjusting rack and the adjusting nut. Then pull the limiting pin, which will shift the adjusting rack to the left side, narrowing the clearance. The adjustment is completed when this clearance becomes nil.

Note: *When the adjusting rack is moved to the extreme left by pulling the limiting pin, the fuel injection volume limiting ring must be so adjusted and set that the A surface of the injection pump and the one-point mark of adjusting rack are in a position of mutual correspondence.*

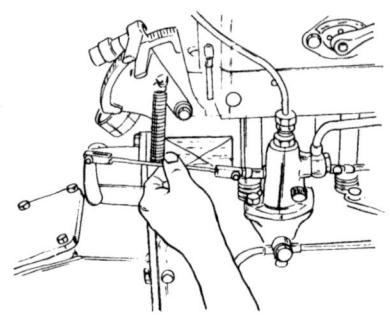

7-3. Fuel Injection Time

Since the fuel injection time has a direct bearing on fuel combustion, if the time is improper, drop of engine brake horsepowers, poor exhaust gas hunting, etc. results.
This engine's proper fuel injection time is 12 degrees before the top dead center (T.D.C.) under the injection volume corresponding to the rated brake horsepowers (injection volume under the condition of the A surface of the injection pump matching with the two-point mark of adjusting rack).
Note: *This engine's fuel injection pump also changes its injection volume with the injection time.*

1. Inspection
    1) Remove the fuel high-pressure pipe and confirm that the delivery port of the injection pump can be seen.
    2) Lower the governor handle and match the two-point mark of adjusting rack with the A surface of the injection pump body.
        Note: *Be sure to adopt the injection volume, corresponding to the position at which the two-point mark matches the A surface, as an injection time checking and adjusting base.*
    3) Turn the flywheel slowly in the normal rotational direction by hand and check on the oil level at the delivery port of the injection pump.
    4) Stop the flywheel at the moment the oil level moves as fuel is delivered.

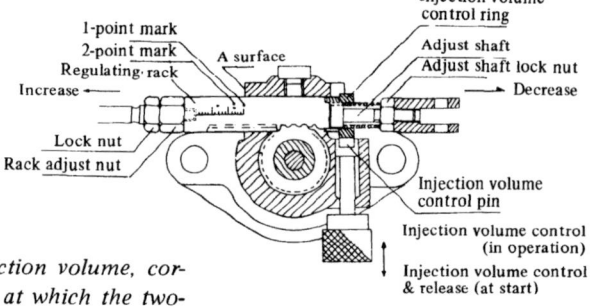

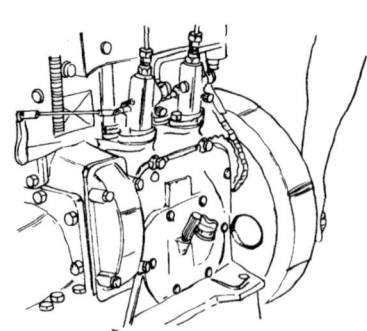

— 69 —

5) At this moment, confirm that the indication on flywheel is 12 degrees before the top dead center (T.D.C.) of compression stroke. If it is, the injection time is proper.
6) If, however, the indication differs from 12° before T.D.C., adjust the time in the following manner:

   Note: *As this engine is 4 cycle, there are two cases of T.D.C. The fuel injection should begin 12 degrees before T.D.C. of compression stroke when both the suction and exhaust valves are closed. The "T" mark on flywheel graduation means the T.D.C.; the "B" mark, bottom dead center (B.D.C.). Figures in front of these marks indicate the cylinder cardinal number; "1" refers to the cylinder closest to the flywheel. The graduation on the flywheel is in 2-degree units.*

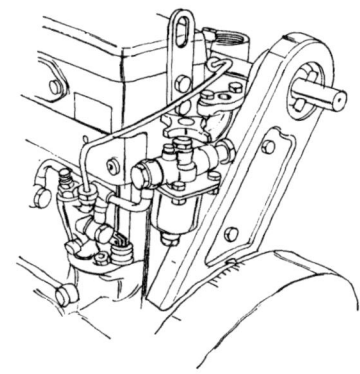

2. Adjustment
   1) Prior to adjustment, it is better to check how much the fuel injection time is advancing or lagging.
   2) Remove the fuel injection pump.
   3) Loose, by using the injection time adjusting spanner, the lock nut of fuel injection time adjusting screw at the roller guide located by the injection pump fitting surface. (Pay attention not to turn the adjusting screw along with the lock nut.)

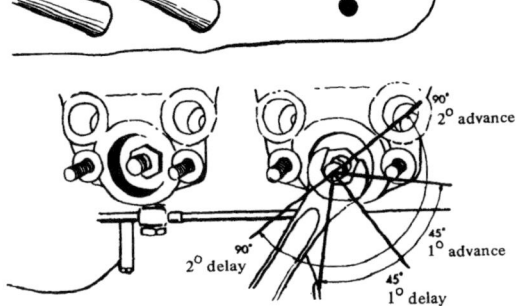

   4) By turning the adjusting screw, adjust the injection time. If the adjusting screw is screwed in, the injection time is delayed. This is shown by the indication on the flywheel approaching T.D.C. On the other hand if the screw is screwed outward, the time is advanced.

      Note: *One complete turn of the adjusting screw changes the injection time by about 7 degrees of flywheel indication. Do not screw out the adjusting screw too far as advancing the injection time too much will cause the pushing up of the injection pump and will result in further trouble.*

   5) Another method is to push down the roller guide with the fingers, turn the flywheel in normal rotational direction by hand, in order to set the indication at 12 degrees before the T.D.C. of compression stroke, and then stop the flywheel from turning further. Next, adjust the adjusting screw so that its upper end matches with the injection time adjusting gauge, a standard accessory, which is located perpendicular to the fitting surface of the injection pump. If this adjustment is done, it results in the adjustment of the fuel injection time to 12 degrees before T.D.C.

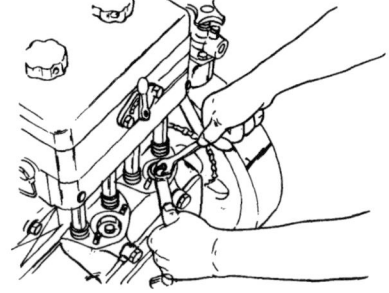

Note: *However, if the injection time still deviates from the proper value by large degrees even after this adjustment is done, it is again necessary to inspect the injection pump.*

6) Once the injection time adjustment is completed, tighten the lock nut on the adjusting screw, using the injection time adjusting spanner.

   Note: *Do not allow the adjusting screw to turn along with the lock nut.*

7) Mount the fuel injection pump on the engine, and re-examine the injection time.

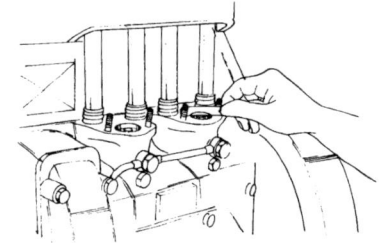

7-4. Fuel Injection Valve

If the fuel to be used contains dust, water and other impurities in large amounts or if it is of an inferior quality, the needle valve is quickly and extensively worn down, resulting in poor injection conditions.

1. Inspection

   For this inspection, use the nozzle tester. In handling the tester, never bring your hands and face near the tip of the injection valve as this is very dangerous.

   1) Attach the injection valve to the nozzle tester, and operate the pump to see at what pressure (kg/cm^2) the fuel injection begins.
   2) If this pressure is 160 kg/cm^2 at the moment of injection, then the pressure is normal.
   3) If not, adjust the injection pressure by the following servicing procedures.

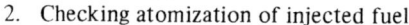

2. Checking atomization of injected fuel

   1) Similar to the injection pressure inspection, check atomization condition of injected fuel from the injection valve mounted to the nozzle tester when the pump is in operation. This checking is also possible by mounting the injection valve through the high pressure pipe to the already inspected and adjusted fuel injection pump attached to the engine in such a way that the injection can be inspected. Then prime the injection pump in order to examine fuel atomization.

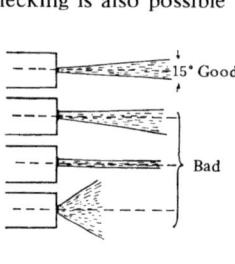

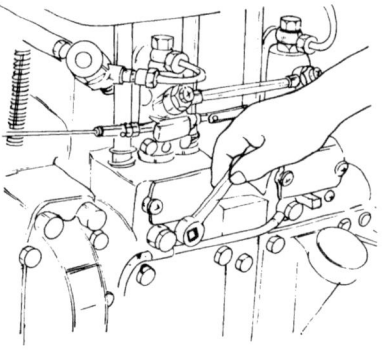

   2) If the atomization of fuel injected from the injection valve forms a conical shape, the injection is proper.
   3) If injected fuel forms a streak or if a considerable quantity of fuel in liquid state leaks out of the injection valve immediately before and after injection or if the injection form is

spread out too much or is too narrow in form or is shifted to one side, the injection is improper.
   4) If the injection is improper, it is necessary to adjust it by the following servicing procedures for the fuel injection valve.
2. Disassembly
   1) If the spring retainer of the fuel injection valve is removed, the pressure adjusting plates, spring, spring holder and inter spindle can be removed.
   2) Take out the fuel strainer tube, and, using the narrow side of the nozzle and the strainer removing tool, remove the strainer.
   3) Screw out the injection valve case nuts, apply the wider side of the nozzle and the strainer removing tool to the precombustion chamber side of the case nuts, and then hammer out the needle valve case.

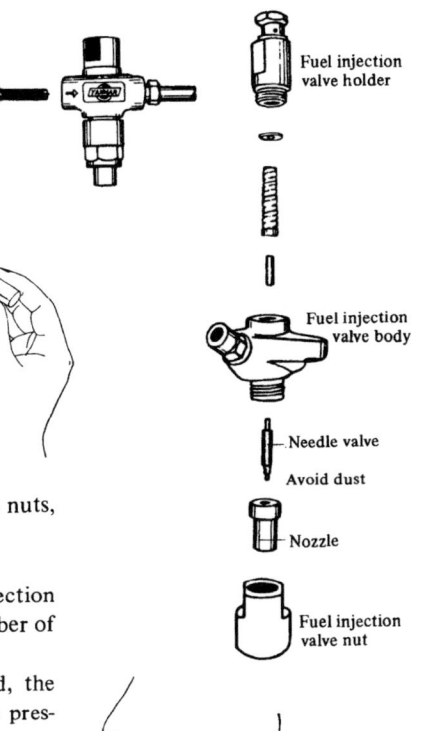

3. Servicing
   1) Adjust the injection pressure of the fuel injection valve by either increasing or decreasing the number of adjusting plates set inside the spring retainer.
   2) If the number of adjusting plates is increased, the injection pressure becomes high; decreased, the pressure drops.
   3) If the needle valve is found to be sticking or has poor contact with the valve case, it is necessary to lap them together. Use fine lapping powder for this purpose, and finish lapping by applying chromium oxide. Replace, with new ones such needle valves and their cases which have no contact between the valve and case at all or whose valve tips have lost luster and become white.
   4) Be sure to replace the needle valve along with its valve case because they are precision machined and lapped together.

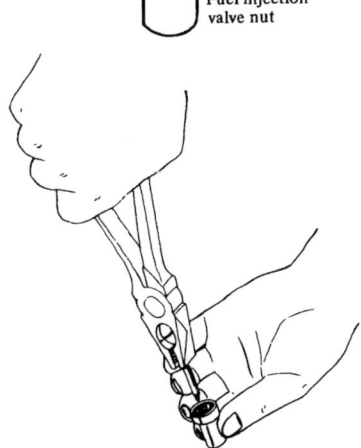

4. Reassembly
   1) Since each component part of the fuel injection valve is processed with precision, it is very important not to scratch or make a scar on any one of these parts when handling. Prior to reassembling these parts, wash them with clean flushing oil, and particularly never allow any dust to enter the needle valve and its case.
   2) Reassemble the fuel injection valve in the opposite order of its disassembly.

7-5. Air venting of fuel injection system

The fuel injection system involves the fuel tank, the fuel injection valve, the fuel strainer, fuel injection pump and fuel high pressure pipe. Air venting can be done using the following procedure.

1) Open the cock of the fuel tank and unfasten the air venting screw of the fuel strainer. Fasten the screw tightly when the fuel, free of bubbles, comes out of this screw.

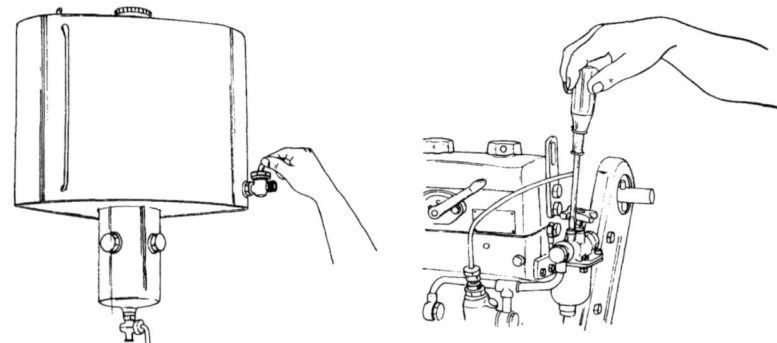

2) Unfasten the air venting screw of the fuel injection pump. Fasten the screw tightly when the fuel, free of bubbles, comes out of this screw.

3) Unfasten the nipple of the fuel injection pump side pipe joint of the fuel high pressure pipe, and set the speed regulator handle at the position of full speed. Then pull the control pin of the fuel injection pump and turn the flywheel several times for priming. This forces the air to come out, along with the fuel, to the outlet of the fuel injection pump. When the fuel, free of bubbles, comes out, tighten the pipe nipple firmly.

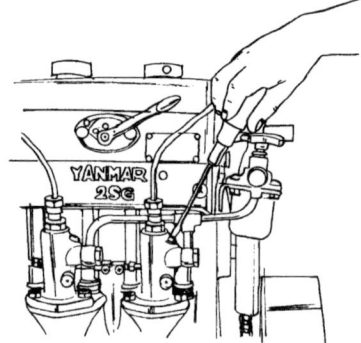

Note: *In case the fuel does not come to the fuel pump after carrying out priming, reaffirm that the fuel comes up to the upper part of the injection pump by the following procedure. Namely, set the control rod of the fuel injection pump temporarily to the flywheel side, setting the regulator handle at the stop position. Then remove the injection valve spring holding the metal of the fuel injection pump.*

4) Set the regulator handle to the full speed position and tighten the fuel high pressure pipe both at the outlet and inlet of the fuel injection pump. After pulling the control pin of the fuel injection pump, turn the flywheel several times by hand. Then the

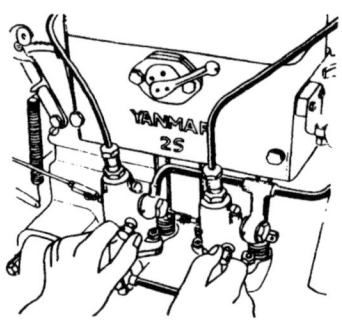

fuel injection sound of "bitz" . . . "bitz" is heard.
5) This means that all the air inside the fuel injection device of one cylinder is completely vented. Once the air venting is carried out, it is not necessary to do it again except when the fuel tank becomes empty or some part of the fuel oil system is removed.

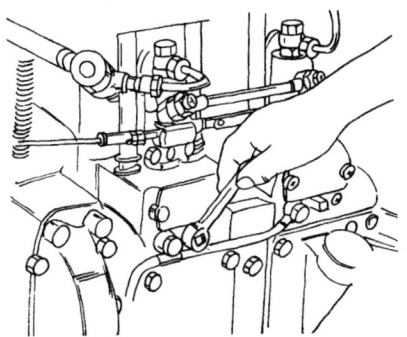

7-6. Cylinder Head

1. Disassembly of the cylinder head assembly
   Refer to the order of engine disassembly for the disassembly of cylinder head. Remove the cylinder head, and clean carbon sticking to the combustion chamber of the cylinder head by flushing it with oil.
2. Disassembly of pre-combustion chamber
   Remove first the fuel injection valve and then dismount the pre-combustion chamber, pushing it out from the fuel chamber side of the cylinder head. Clean the inside of the injection nozzle well, because, if the engine is operated for a long duration under imperfect combustion conditions, carbon may be deposited inside the pre-combustion chamber. At the same time, check the copper packing to see that it is not scratched, and if so, replace such copper packing with new packing in order to avoid compression leakage.
3. Disassembly of suction and exhaust valves
   1) Hold down the spring using the spring inserting and removing tool (standard attachment) and then take out the holding metal of the spring shoe.

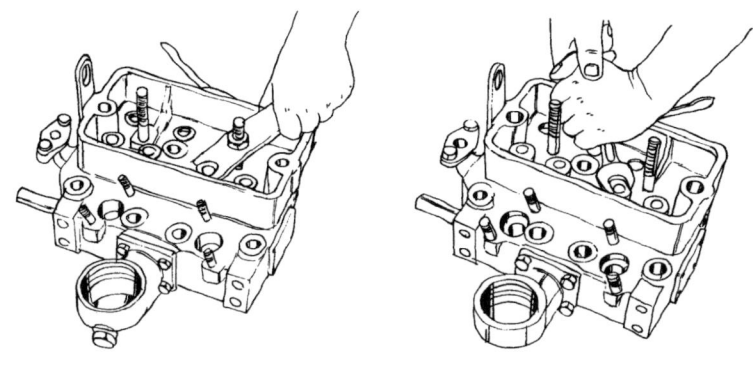

2) Remove the circlip retaining the valve.
3) Remove the valve.

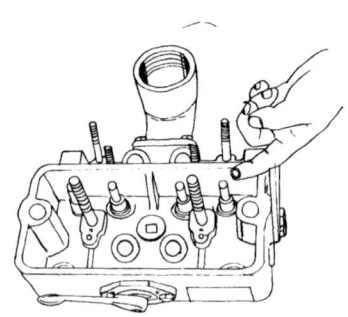
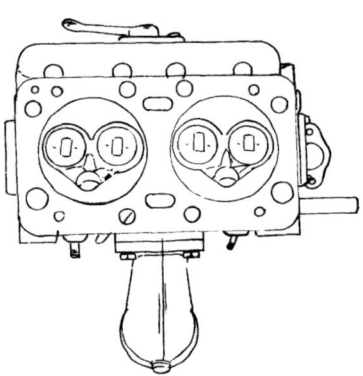

4. Lapping of suction and exhaust valves
   1) After cleaning up the valve seat of the cylinder head, apply a little talc on the valve seat and place the valve thereon. Wriggle the valve up and down several times in order to find out the contact point. Lapping is to be done using the following procedure for those valves in bad contacting condition.
   2) In case the contact of the valve is not satisfactory (having irregular surfaces or excessive clearance), cut off the valve seat of the cylinder head as little as possible with a valve seat cutter. Valve seats having more than about 2mm thickness should also be cut with a cutter.

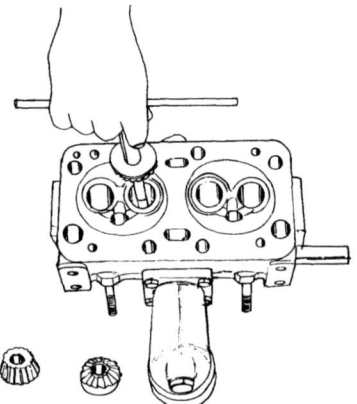

Specification of Seat Cutter				
Outside dia. of cutter	Angle for seat	Angle for outer side	Angle for inner side	Rod dia. mm
45φ	45°	15°	65°	10φ

   3) Using the valve lapping tool and lapping powder (both are standard attachments), the contact point is made on the valve seat using rough lapping powder first and then fine lapping powder. Finally, lapping should be done with lubricating oil so that the contact of the seat surface becomes uniform at every point.
   Note: *Lapping should be done by tapping the valve seat lightly while turning it little by little, rather than turning it around often.*

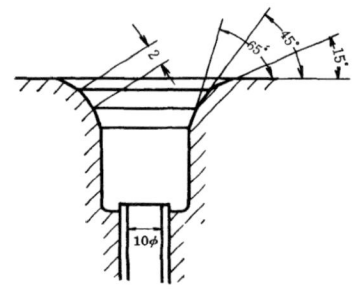

4) Repeating the instructions in 1), the contacting condition may be checked again. If the contact is circumferential and continuous, the lapping has been done satisfactorily.

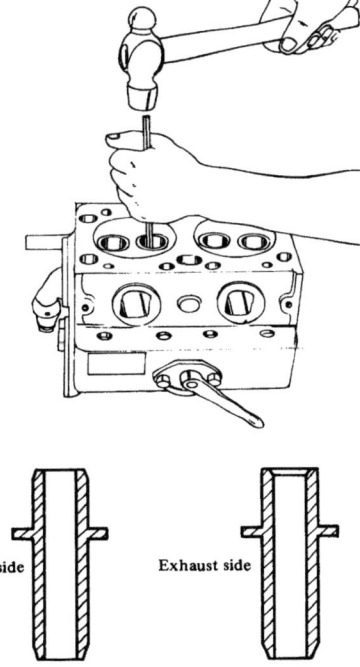

5. Replacement of suction and exhaust valve guide
   The replacement of suction and exhaust valve guides can be done using the following procedure.
   1) Remove the suction and exhaust valves.
   2) To remove the valve guide, place a brass or copper rod with the same diameter as the valve guide at the valve guide edge of the cylinder head combustion chamber, and hammer it out.
   3) To replace the valve guide, using a hammer, insert the new valve guide from the valve lever case side attaching a pad to prevent the valve guide from being scratched. In this case, clean up the hole of the cylinder head and put lubricating oil on it beforehand.

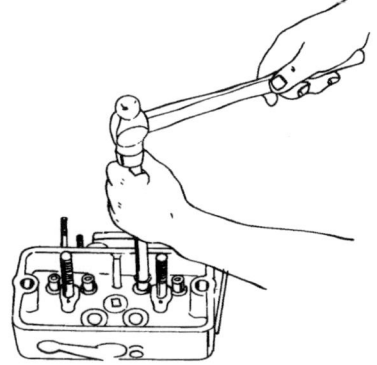

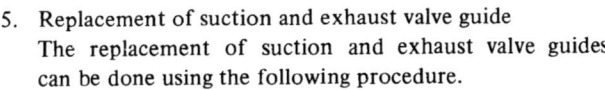

Suction side   Exhaust side

Note: *Since there is a difference between the suction side and exhaust side, making a mistake should be avoided, when they are replaced.*

4) Check that the suction and exhaust valves slide smoothly when they are inserted into the guide.
5) Check the contact of the valve seat, and give a proper lapping for any valve seats with inferior contact.

6. Assembly of cylinder head
    The assembly of cylinder head can be done according by reversing the order of the disassembly. Pay attention to the following points.
    1) Clean fully the lapping powder which may remain after lapping the valve.
    2) Do not take the lapped valve for the valve seat.
    3) Apply lubricating oil to the valve rod and the sliding part of the valve guide, and the valve seat.

7. Installation of cylinder head assembly
    Refer to the order of engine assembly for the installation of the cylinder head assembly. It is necessary to adjust the opening and closing time of the suction and exhaust valves after the cylinder head is installed. At the same time, check the top clearance as it may change when the copper packings of the cylinder head assembly are replaced. The checking is to be done according to the following procedure.

8. Top clearance checking procedure
    1) Put lead of around $3\phi \times 5$ on three flat places on top of the piston, and mount the cylinder head (lock torque: 12 kg/m). Then turn the crank shaft slowly once by hand.
    2) Remove the cylinder head and take out the lead being placed on the piston. Measure the thickness (top clearance) of the pressed lead.
    The standard top clearance is 1.85 ± 0.1 mm.
    3) The top clearance can be adjusted by means of adjusting the thickness of the gasket packing.

9. Inspection and Adjustment of the Opening and Closing Times of the Suction and Exhaust Valves
    There are two methods inspecting and adjusting the opening and closing times of the suction and exhaust valves. In either case, perform the inspection and adjustment when the engine is in a cooled down state.

    • Method 1: "Valve Head Clearance of Suction and Exhaust Valves" Method
    1) When the engine is at the T.D.C. of the compression stroke, in which both the suction and exhaust valves are completely closed, check the clearance between the valve lever and valve head by using the thickness gauge, a standard accessory, to see that the clearance is 0.2mm for all of the cylinders.
        Note: *The compression handle should be at its "RUN" position for this checking purpose.*

    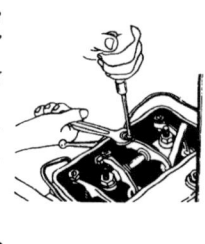
    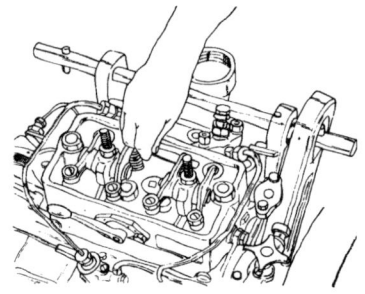

    2) The adjustment is to be done by loosening the lock nut from the adjusting screw, and setting the screw in a position yielding the proper valve head clearance of 0.2mm.

3) After the adjustment, tighten up the lock nut.
- Method 2: "Flywheel Graduation" Method
1) For suction valve of each cylinder,
   a. While turning the flywheel in the normal rotational direction by hand, watch the motion of the suction valve lever or push rod.
   b. Stop the flywheel from turning at the moment the valve lever begins to push the suction valve, and take a reading of the flywheel graduation.
   c. If this indication is the "I.O." position before the T.D.C., one graduation from the T.D.C. of compression stroke, then everything is in order.
   d. If not, first bring the flywheel to the position at which the pointer indicates the "I.O." mark the flywheel graduation, loose the lock nut from the adjusting screw, and adjust the screw so that the valve lever is brough in contact with the valve. After that, tighten up the lock nut.
   e. Again turn the flywheel in the engine's normal rotational direction until the valve lever begins to separate from the valve. At this moment if the flywheel graduation indicates the "I.C." mark, then everything is in order.
2) For exhaust valve of each cylinder,
   a. While turning the flywheel in the normal rotational direction by hand, water the motionof the exhaust valve lever or its push rod.
   b. Stop the flywheel from turning at the moment the valve lever begins to separate from the valve, and take a reading of the flywheel graduation.
   c. If this indication is in the "E.C." position before the T.D.C., one graduation from the T.D.C. of the compression stroke, then everything is in order.
   d. If not, first bring the flywheel to the position at which the pointer indicates the "E.C." mark on the flywheel graduation, loose the lock nut from the adjusting screw and adjust screw so that the valve lever is brought in contact with the valve. After that, tighten up the lock nut.
   e. Again turn the flywheel in the normal rotational direction until the valve lever begins to push on the valve. At this moment if the flywheel graduation indicates the "E.O." mark, then every thing is in order.

Note: *Although the decompression handle may be set at the decompression position when inspecting and adjusting the suction valve, be sure to set the handle to the "RUN" position when performing the inspection and adjustment on the exhaust valve. The decompression handle may be operated to bring down the pressure if compression pressure is applied during the manual turning of the flywheel.*

Reference: Since this engine is 4-cycle, there are two cases of top dead center. The suction valve opening position and the exhaust valve closing position are 18 degrees before and 15 degrees after the top dead center (T.D.C.) exactly one complete turn from the T.D.C. of the compression stroke, respectively. The "I.O." mark on

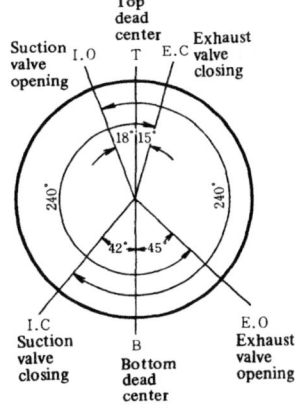

the flywheel graduation refers to the suction valve opening; the "I.C.," the suction valve closing. The "E.O." mark refers to the exhaust valve opening the "E.C.," to the exhaust valve closing. Furthermore, the "T" mark indicates the top dead center (T.D.C.); the "B" mark the bottom dead center (B.D.C.). The figures preceeding the above mentioned marks refer to cardinal numbers for the cylinders "1" refers to the cylinder closest to the flywheel side, "2" the second cylinder, "3" the third cylinder, etc. Therefore, the "IT" mark means the T.D.C. of the first cylinder.

7-7. Piston Pin and Connecting Rod

1. Removing the piston and connecting rod assembly
   As for removing the piston and connecting rod assembly refer to Section "Procedure on Disassembling the Engine."
2. Disassembling of Piston Ring
   Remove the ring by the use of two thin wire rings (about 50mmϕ) as shown in figure.
   Note: *Be careful not to break the wires due to excessive widening.*
3. The piston ring is composed of the following:
   Compression ring of 3.5mm in width: 1 pc
   Compression ring of 2.5mm in width: 2 pcs
   Oil scraper ring of 4mm in width: 2 pcs
4. Removing the piston pin
   1) Remove the piston pin fixing rings, which are fitted to both ends of piston pin holes, by using pliers.
   2) Soak the piston pin in lubricating oil heated to nearly 80°C and warm it for about 10 minutes. Apply a rod to the piston pin and tap it lightly with a hammer to remove the piston pin.
      Note: *As the piston pin fitting part has a small fitting allowance at the temperature near the ordinary temperature of the piston it is necessary to warm the piston to expand it and then remove the piston pin.*

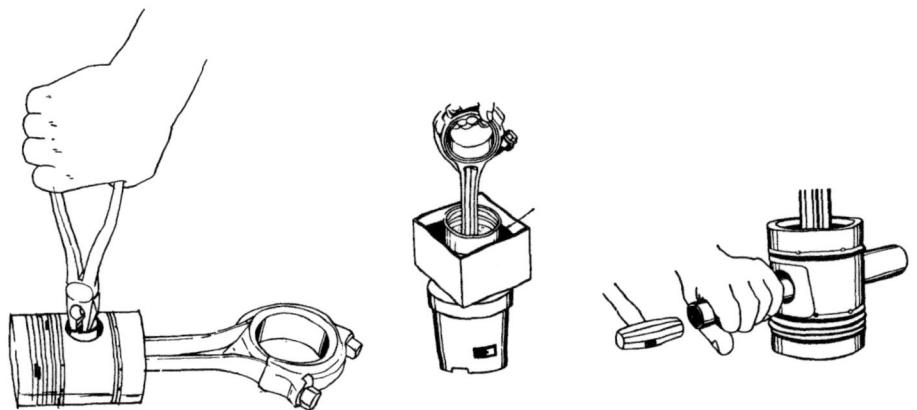

5. Service
   1) Replace the piston and piston ring if necessary according to "Maintenance Standards of Main Parts."
   2) After about 1000 hrs operation the piston ring should be changed with a new one as it will probably be remarkably worn and its tension reduced.
   3) Replace the piston pin, the piston pin metal and the crank pin metal according to "Maintenance Standards of Main Parts," if necessary. The crank pin metal can not be adjusted by means of shim (metal liner).
   4) The under size metal can be used according to the degree of wear of the crank pin.
6. Assembling the piston ring
   1) Clean the piston head; the using grooves and the oil holes throughly before setting the piston ring.
   2) Set the piston ring with the marked side, which is stamped on near the notched part of piston ring up.
   3) Do not use the plated ring since the inner surface of the cylinder liner is chrome-plated.
7. Assembling the piston pin
   1) The assembling of the piston pin is accomplished by reversing the procedure on "Removing the Piston Pin" above.
   2) As for mounting the piston and connecting rod assembly, refer to the Section, "Procedure on Reassembling."

7-8. Cylinder Liner

1. Removing the cylinder liner
   Remove the cylinder head assembly and the piston connecting rod assembly.
   1) Refer to the Section, "Procedure on Disassembling."
   2) On taking out the liner use the liner removing tool (refer Fig.).
   3) Fit the disc plate (1) to the under part of the liner and insert the nut into the cylinder head, tightening bolt, thus making a support (or use another support).
   4) Insert the washer (2), tighten the disc (1) to the washer (2) by means of bolt (3) and then gradually clamp the upper nut (4) thus making it possible to remove the liner.

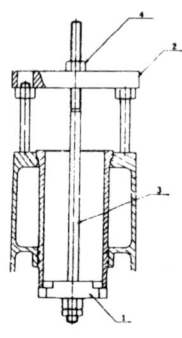

2. Inserting of cylinder liner
   1) Before inserting the cylinder liner clean the dust and the paint from the liner fitting hole cylinder block.
   2) Insert a new rubber packing into the groove of the liner with care avoiding the distortion of the packing.
   3) Apply the white paint diluted by lubricating oil to the rubber packing section and then insert the liner with care tighten it at the cylinder bed once.
   4) Check the inner diameter with a cylinder gauge. The allowable range of distortion is be within 0.02mm.

Rubber packin

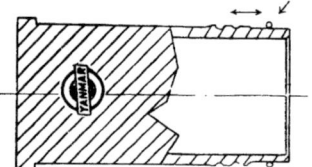

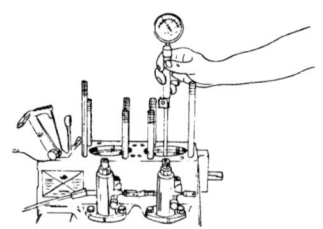

7-9. Replacement of Crank Journal Metal

Under-size and over-size metals are specially provided to be used according to the extent of abrasion of the inside part and the thrust part of the crank journal metal. Refer to "Maintenance Standards of Main Parts." Replace inferior metals if any.

1. Removel of metal
   1) Remove the set bolts of metal.

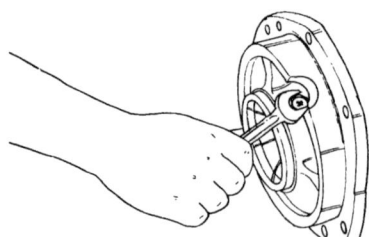
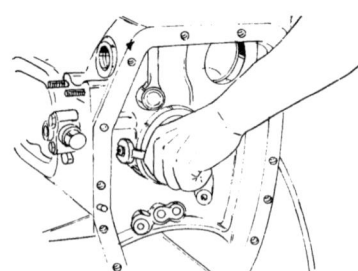

   2) Take out the metal by holding it down from the side without a flange.

— 81 —

2. Metal Setting
   1) The metals are set by pressing it so that the oil hole of the flywheel side metal comes to the oil hole of the gear side metal to the oil hole of the cylinder main.
   2) Attach the set bolts and tighten the lock nuts
      Note: *Be sure to put lubricating oil on the outside surface of the metal when it is set. Do not make mistakes in setting the position of the oil hole and the set bolts hole. Avoid deforming the metal in its setting. Measure the inside diameter of metal after it has been set.*
      Note: *The inside clearance of the crank shaft (inside clearance between crank journal and its metal) should be adjusted to 0.15 ~ 0.3mm by means of increasing or decreasing the number of metal housing packings.*

7-10. Cam Shaft and Cam Shaft Mountings

1. Removing
   1) As for removing refer to the Section "Procedure on Disassembling."
   2) Minimum procedure on removing the cam shaft assembly only is as follows.
      a. Remove the reverse gear upper cover.
      b. Remove the flywhee. (Refer to "Procedure on Disassembling.")
      c. Take off the chain.
      d. Take away the valve lever shaft support.
      e. Remove the push rod.
      f. Remove the crank case side cover plate at cam shaft side only.
      g. Remove both fuel pump and roller guide. (Refer to "Procedure on Disassembling.")
      h. After that take out the cam shaft. As for subsequent procedures refer to the Section "Procedure on Disassembling."
2. Mounting
   1) Mounting is carried out by reversing the Procedure on Removing according to "Procedure on Reassembling."

7-11. Cooling Water Pump

As the application of cooling water, which contains large quantities of mud, sand and dirt, results in the clogging of suction and exhaust valves as well as the wearing of the valve seats, the contact becomes faulty and the amount of discharged water decreases. Therefore, it is recommended that cooling water as pure as possible be used.

1. Checking (on the amount of discharged cooling water)
   1) Measure the amount of water discharged out of the outlet at nominal revolution rate of engine (1800 rpm). The standard delivery is as follows.
      544 ℓ/1800 rpm-1 hr
   2) When delivery is lacking it is necessary to carry out checking as follows.
      a. Check that the cooling water inlet (Kingstone) is not clogged.
      b. Check that the inlet cock (Kingstone cock) is fully opened.
      c. Check that there is no hole on the cooling water pipe or that the air is sucking in from the packing part of the flange.
      d. Check that the cooling water pipe is not clogged.
      e. Check that there is no dust in the suction and exhaust valves

of the pump.
  f. Check the contact of suction and exhaust valves to each valve seat.
  g. Check the sliding between the valve and the valve guide.
  h. Check that the valve spring is not weakened.
  i. Check the tightness of gland packing.
2. Removing cooling water pump
   As for removing the cooling water pump from the engine refer to the Section "Procedure on Disassembling."
3. Dismantling
   1) The valve spring and the suction valve can be removed by taking off the suction valve guide (with a packing).
   2) After removing the air chamber (with a packing) and the exhaust valve guide (with packing), the valve spring and the exhaust valve can be taken down.

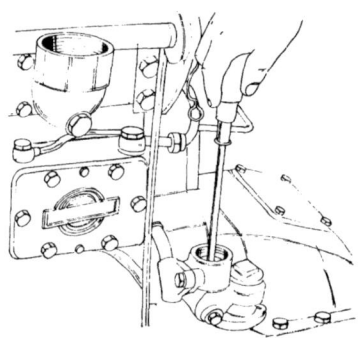

4. Maintenance
   Adjust the contact of suction and exhaust valves by lapping with the use of fine lapping powder.
5. Reassembling
   Reassembling is carried out by reversing the dismantling procedure.
   After lapping the suction and exhaust valves, clean the valves and the valve seats so that no lapping powder remains in the pump and apply lubricating oil to the valve seats. As the air tightness of the valve is secured no pre-charging water is required.
6. Assembling cooling water pump
   To mount the cooling water pump on the engine refer to the section "Procedure on Reassembling."
7. Check the delivery of pump by driving the engine.

7-12. Lubricating Oil Pump and Lubricating Oil Pressure Adjusting Valve

1. Lubricating oil pressure check
   1) Watch the oil pressure gauge to see that the lubricating oil is fully warmed up, operating the engine at the standard number of rotation (1800 rpm).

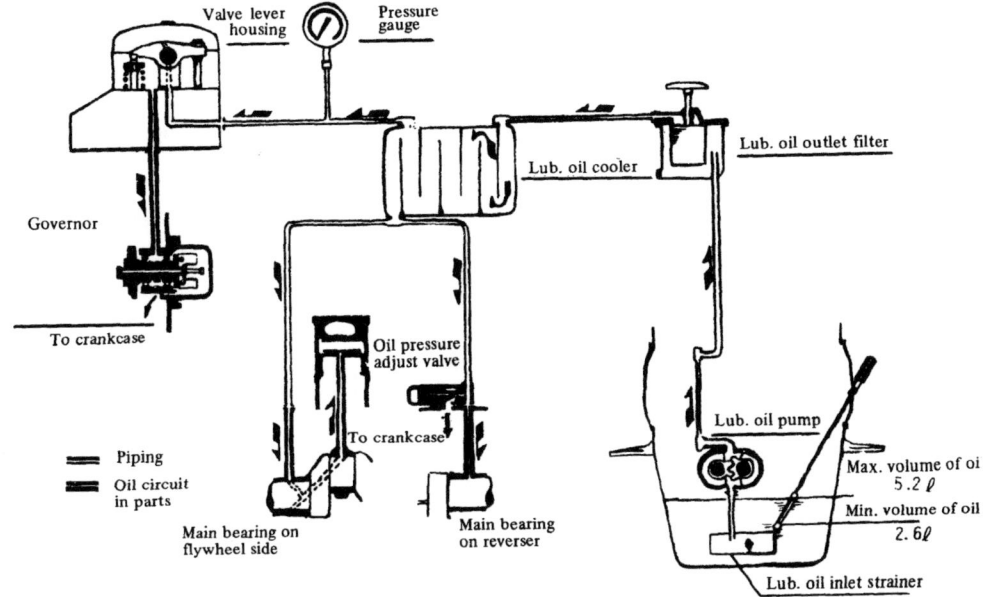

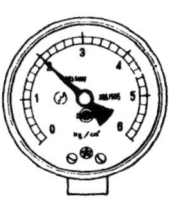

2) It is in a normal condition if the hand of the oil pressure gague shows the position of $2 \sim 2.5$ kg/cm².
3) In case the hand is showing a position other than the above, it is necessary to adjust the oil pressure adjusting valve.
   Note: *The oil pressure may be over 4 kg/cm² soon after the engine is started in cold weather, and it may sometimes be lower than 1 kg/cm² when the engine is operated slowly. But in any case the oil pressure adjusting valve should always be adjusted so that it comes within $2 \sim 2.5$ kg/cm² in the above state.*

2. Adjusting lubricating oil pressure
   1) Remove the cap ① (with copper packing) of the oil pressure adjusting valve attached to the lubricating oil strainer.
   2) Adjust the spring strength which holds the valve, by means of turning the spring holder ③ after releasing the lock nut ②.

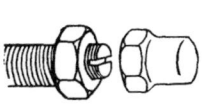

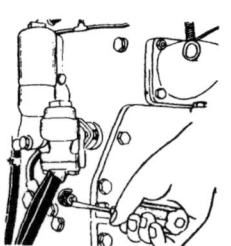

3) The oil pressure is increased when the spring holder ③ is turned clockwise, and lowered when turned counter clockwise.
4) Adjust the lubricating oil pressure so that the hand of the oil pressure gauge comes to the position of $2 \sim 2.5$ kg/cm² when the lubricating oil is fully warmed up, operating the engine at the standard number of rotations (1800 rpm).
5) When the adjustment is finished, attach the cap ① (with copper packing.) after setting the lock nut ②.

3. Checking lubricating oil pump
    1) Remove the cap ① of oil pressure adjusting valve when the lubricating oil is fully warmed up, operating the engine at the standard number of rotations (1800 rpm).
    2) Watch the graduation of the oil pressure gauge pressing the valve head ④ of the oil pressure adjusting valve.
    3) The pump is in a normal condition if the pressure is within $2 \sim 2.5$ kg/cm².
        Note: *Most of the oil discharged from the lubricating oil pump escapes through the oil pressure adjusting valve. If conditions are bad, the lubricating oil pressure is not increased due to a too great escape of oil because of the malfunctioning of the oil pressure adjusting valve. In this case, it is necessary to check without fail the condition of the oil pressure adjusting valve.*

4. Disassembing lubricating oil pump
    1) Refer to the order of engine disassembly for disassembling the lubricating oil pump from the engine.

5. Disassembling lubricating oil pump
    1) Unfasten the driving gear lock nuts (with split pin).
    2) Dismount the driving gear using a dismounting tool. (Inserting of taper and key.)

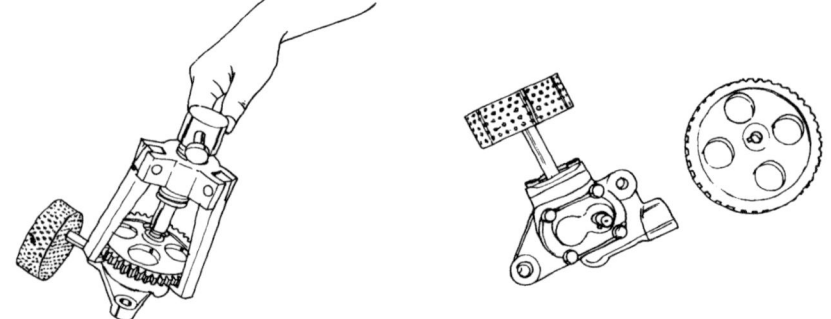

    3) Take out the driving gear key.
    4) Remove the strainer net unfastening the strainer net fitting bolts (with wire).
    5) Removing the bolt (with wire), each gear inside the pump can be removed.
    6) Clean lubricating oil pump.

6. Servcing of lubricating oil pump
    1) If the clearance between the shaft and the case Ⓐ and between the gear side face and the gear (driving and being driven) inside the pump is excessive, the pump efficiency is decreased causing a shortage of discharged oil.
    2) Such a pump, incapable of discharging a sufficient amount of oil and having excessive clearance as mentioned above, should be replaced.

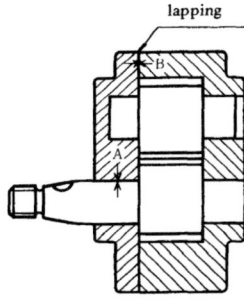

	Standard Clearance (min. ~ max.) mm
A	0.014 ~ 0.056
B	0.014 ~ 0.056
Standard amount	327 ℓ/1800 rpm-hr

7. Assembling
   1) The assembling can be done by means of reversing the order of the disassembling.
   2) Lapping has been done on the joint surface of the case, so that scratching the surface or leaving dust on it should be avoided.
8. Assembling lubricating oil pump
   1) Refer to the order of engine reassembly for the reassembling lubricating oil pump engine.
   2) The condition of the driving shaft rotation of the pump should be checked in reassembling. Reassemble the main body of the pump so that the back lash between the crank gear and pump driving gear comes within the range of 0.08 ~0.16mm.
9. Adjustment of reversing gear
   The reversing gear built in this engine does not require any readjustment.

8. COUNTERMEASURES TO ENGINE TROUBLES

Sympton	Check Point	Trouble			Treatment	Reference
1. Hard to crank up	Lub. oil	Extremely high viscosity			Replace	Refer to page 17.
		Deterioration				
	Moving parts	Seizure of piston			Repair or replace	
		Seizure of crank journal metal				
		Seizure of crank pin metal				
		Seizure of cam shaft metal				
		Seizure of reversing gear box metal				
		Extremely small clearance of metal			Adjust	
		Seizure of each guide and sliding part			Repair or replace	
		Improper centering of shaft or pump			Adjust	
2. Leakage of compressed air	Suction & exhaust valves	Clearance of valve seat			Repair or adjust lapping the valve	Refer to page 75.
		Clogging of valve seat			Clean	
		Improper valve timing			Adjust	Refer to page 77.
	Cylinder head	Insufficient tightness			Retighten	Refer to page 77.
		Disorder of copper packing			Replace	
	Fuel injection valve	Improper tightness			Retighten	
		Disorder of copper packing			Replace	
	Piston ring	Wear			Replace	Refer to page 80.
		Sticking			Replace	
		Cuts of piston ring lined up in one line			Adjust the rings or check the piston liner	
	Cylinder liner	Wear			Replace	Refer to page 81.
		Separation of plating layer			Replace	
3. No injection of fuel oil	Fuel tank	Lack of fuel oil			Supply fuel oil	
	Fuel cock	Remaining closed			Open	
	Governor handle	Remaining unraised			Raise	
	Fuel pipe	Clogging			Clean	
	Fuel system	Air mixed with fuel			Vent air	Refer to page 73.
	Fuel injection valve	Wear			Replace	
		Sticking			Lap or replace	
		Broken of spring			Replace	
		Loosened case nuts or spring retainer			Tighten	
	Fuel injection pump	Exh. valve	Clogging		Clean	Refer to page 65.
			Disorder of valve seat		Lap or replace	
			Broken spring		Replace	
		Plunger	Wear			
			Seizure		Replace	
			Broken spring			
			Loosened plunger barrel set screw		Tighten	
		Improper pump assembly matchmark			Adjust	
	Fuel regulating device	Improper regulation			Adjust	Refer to page 67.

Sympton	Check Point	Trouble	Treatment	Reference
4. No fire even upon fuel injection	Fuel injection valve	Improper injecting pressure	Adjust	Refer to page 71.
		Improper injection	Lap or replace	
	Fuel injection time	Improper regulation	Adjust	Refer to page 69.
	Fuel oil	Improper fuel oil	Replace	Refer to page 15.
		Water mixed in fuel	Replace	
	Combustion chamber	Carbon clogging	Clean	
	Insufficient compressing	Leakage of air compressed		Refer to 2.
5. Hard to start	Difficulty in revolution			Refer to 1.
	Insufficient ignition			Refer to 3 & 4.
6. Gradual stop of engine	Fuel system			Refer to 3 & 4.
	Cooling system	Overheating due to lack of lub. oil	Supply lub. oil	
		Overheating due to lack of cooling water	Check the cooling water pump or lap the valve	Refer to page 82.
	Overload		Reduce load	
7. Insufficient output	Insufficient compression pressure			
	Irregular combustion			
	Propeller	Incorrect matching	Select propeller	
	Moving parts	Overheating or seizure	Repair or replace	Refer to 1.
8. Bad colored exhaust gas	Lub. oil	Oversupply	Supply proper amount of oil	Refer to page 16.
	Piston ring oil ring	Wear or breakage	Replace	Refer to page 82.
	Overload		Adjust load	
	Lack of output			Refer to 7.
	Fuel injection time	Too early or delayed injection	Adjust	Refer to page 69.
	Combustion chamber	Carbon sticking or melt down	Clean or replace	
9. Irregular revolution (hunting & over running)	Fuel regulating device	Incorrect regulation	Adjust	Refer to page 67.
		Difficulty in regulating lever motion	Repair	
	Governor	Faulty function	Repair	
	Fuel system			Refer to 3 & 4.
10. Abnormal engine noise (knocking)	Fuel injection valve	Extremely high injection pressure	Adjust	Refer to page 71.
		Faulty injection	Lap the valve or replace it	
	Fuel injection time	Too early or delayed injection	Adjust	Refer to page 69.
	Flywheel	Loosened	Tighten	
		Loosened parts	Tighten	
		Increase of metal clearance	Adjust	
	Fuel oil	Water mixed in fuel	Replace	
	Cooling water pump	Overheat of engine	Check the cooling water pump or lap the valve	
		Incorrect centering of shaft	Center	
	Bearing	Wear	Replace	

9. STORING ENGINE

1. When the engine is used in cold districts it is necessary to drain the cooling water of the engine (including cooling water pump).
   Note: *When opening the cooling water cock the water in the engine is discharged. By taking out the plug, the water in the cooling water pump is released.)*

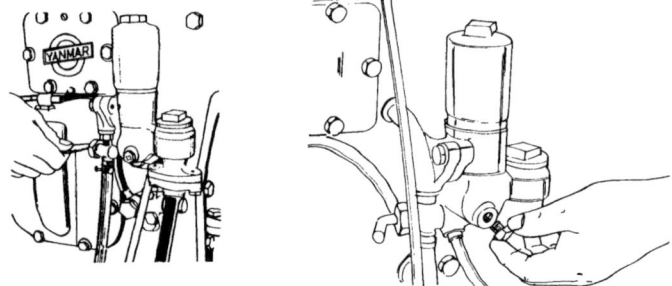

2. If rainwater may enter into the machine from the exhaust pipe, cover it.
3. Set the suction and exhaust valves of No. 1 and No. 2 cylinders to closed state and then stop the flywheel.
   Note: *In order to obtain the closed state of suction and exhaust valves as for No. 1 and No. 2 cylinders set the No. 1 cylinder to position passing through its compression upper dead point by 90° (where the position of flywheel start handle is coincident with the pointer) and set the decompression handle to the operating state.*

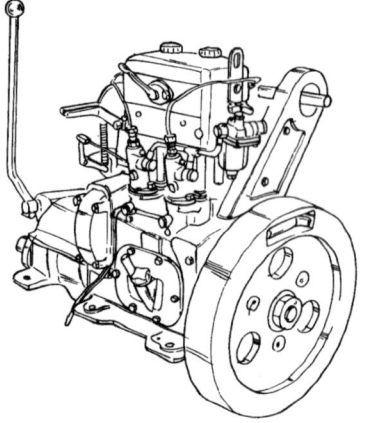